Foseco Ferrous Foundryman's Handbook

Foseco Ferrous Foundryman's Handbook

Edited by
John R. Brown

OXFORD   AUCKLAND   BOSTON   JOHANNESBURG   MELBOURNE   NEW DELHI

Butterworth-Heinemann
Linacre House, Jordan Hill, Oxford OX2 8DP
225 Wildwood Avenue, Woburn, MA 01801-2041
A division of Reed Educational and Professional Publishing Ltd

 A member of the Reed Elsevier plc group

First published 2000

British Library Cataloguing in Publication Data
Brown, John R.
    Foseco ferrous foundryman's handbook.
    1. Founding – Handbooks, manuals, etc.
    I Title   II Foseco International   III Foundryman's handbook
    671.2

Library of Congress Cataloguing in Publication Data
Foseco ferrous foundryman's handbook/revised and edited by John R. Brown.
    p. cm.
    Includes index.
    ISBN 0 7506 4284 X
    1 Founding – Handbooks, manuals, etc.   2 Iron founding – Handbooks,
    manuals, etc.   3 Cast-iron – Handbooks, manuals, etc.   I Title: Ferrous
    foundryman's handbook   II Brown John R.

    TS235 F35 2000
    672.2–dc21   00-033679

ISBN 0 7506 4284 X

Typeset at Replika Press Pvt Ltd, Delhi 110 040, India
Printed and bound in Great Britain by Biddles Ltd, *www.biddles.co.uk*

Contents

Preface

The last edition of the *Foseco Foundryman's Handbook* was published in 1994 and like all the earlier editions, it aimed to provide a practical reference book for all those involved in making castings in any of the commonly used alloys by any of the usual moulding methods. In order to keep the *Handbook* to a reasonable size, it was not possible to deal with all the common casting alloys in detail. Since 1994 the technology of casting has continued to develop and has become more specialised so that it has been decided to publish the new edition of the *Handbook* in two volumes:

Ferrous   dealing with grey, ductile and special purpose cast irons
              together with carbon, low alloy and high alloy steels
Non-ferrous   dealing with aluminium, copper and magnesium casting
              alloys

Certain chapters (with slight modifications) are common to both volumes: these chapters include tables and general data, sands and sand bonding systems, resin bonded sand, sodium silicate bonded sand and feeding systems. The remaining chapters have been written specifically for each volume.

The *Handbook* refers to many Foseco products. Not all of the products are available in every country and in a few cases, product names may vary. Users should always contact their local Foseco company to check whether a particular product or its equivalent is available.

The Foseco logo and all product names appearing in capital letters are trademarks of the Foseco group of companies, used under licence.

John R. Brown

Acknowledgements

The following Organisations have generously permitted the use of their material in this book:

The American Foundryman's Society Inc., 505 State Street, Des Plaines, Illinois 60016-8399, USA

British Standards Institution (BSI), Extracts from British Standards are reproduced with the permission of BSI. Complete copies can be obtained by post from BSI Customer Services, 389 Chiswick High Road, London W4 4AL

Butterworth-Heinemann, Linacre House, Jordan Hill, Oxford OX2 8DP

The Castings Development Centre (incorporating BCIRA), Bordesley Hall, The Holloway, Alvechurch, Birmingham, B48 7QB

The Castings Development Centre (incorporating Steel Castings Research & Trade Association), 7 East Bank Road, Sheffield, S2 3PT

The Department of the Environment, Transport and the Regions

Foundry Management & Technology, Penton Media Inc., 1100 Superior Avenue, Cleveland OH 44114-2543, USA

Foundry & Technical Liaison Ltd., 6–11 Riley Street, Willenhall, West Midlands, WV13 1RV

Institute of British Foundrymen, Bordesley Hall, The Holloway, Alvechurch, Birmingham, B48 7QA

Rio Tinto Iron & Titanium GmbH, Eschborn, Germany

SinterCast Ltd, Regal House, 70 London Road, Twickenham, Middlesex

The author gratefully acknowledges the help received from colleagues at Foseco International Limited, Foseco UK Limited and other Foseco Companies.

Chapter 1
Tables and general data

SI units and their relation to other units

The International System of Units (SI System) is based on six primary units:

Quantity	Unit	Symbol
length	metre	m
mass	kilogram	kg
time	second	s
electric current	ampere	A
temperature	degree Kelvin	K
luminous intensity	candela	cd

Multiples

SI prefixes are used to indicate multiples and submultiples such as 10^6 or 10^{-3}

	Prefix	Symbol		Prefix	Symbol
10	deca	da	10^{-1}	deci	d
10^2	hecto	h	10^{-2}	centi	c
10^3	kilo	k	10^{-3}	milli	m
10^6	mega	M	10^{-6}	micro	μ
10^9	giga	G	10^{-9}	nano	n
10^{12}	tera	T	10^{-12}	pico	p

Example: One millionth of a metre is expressed as one micrometre, 1 μm.

Derived units

The most important derived units for the foundryman are:

Quantity	Unit	Symbol
Force	newton	N (kg m/s^2)
Pressure, stress	newton per square metre or pascal	N/m^2 (Pa)
Work, energy	joule	J (Nm)
Power, heat flow rate	watt, joule per second	W (J/s)
Temperature	degree Celsius	°C
Heat flow rate	watt per square metre	W/m^2
Thermal conductivity	watt per metre degree	W/m K
Specific heat capacity	joule per kilogram degree	J/kg K
Specific latent heat	joule per kilogram	J/kg

SI, metric, non-SI and non-metric conversions

Length:
1 in   = 25.4 mm
1 ft   = 0.3048 m
1 m   = 1.09361 yd
1 km   = 1093.61 yd = 0.621371 miles
1 mile = 1.60934 km = 1760 yd
1 yd   = 0.9144 m

Area:
1 in^2 = 654.16 mm^2
1 ft^2 = 0.092903 m^2
1 m^2 = 1.19599 yd^2 = 10.76391 ft^2
1 mm^2 = 0.00155 in^2
1 yd^2 = 0.836127 m^2
1 acre   = 4840 yd^2 = 4046.86 m^2 = 0.404686 m^2 hectare
1 hectare = 2.47105 acre = 10 000 m^2

Volume:
1 cm^3 = 0.061024 in^3
1 dm^3 = 11 (litre) = 0.035315 ft^3
1 ft^3 = 0.028317 m^3 = 6.22883 gal (imp)
1 gal (imp) = 4.54609 l (litre)
1 in^3 = 16.3871 cm^3
1 l (litre)   = 1 dm^3 = 0.001 m^3 = 0.21997 gal (imp)
1 m^3 = 1.30795 yd^3 = 35.31467 ft^3
1 pt (pint)   = 0.568261 l
1 US gal   = 3.78541 l = 0.832674 gal (imp)

1 ft^3/min (cfm) = 1.699 m^3/h
1 ft^3/sec   = 28.31681/s

Mass:
1 lb (pound) = 0.453592 kg
1 cwt   = 50.802 kg
1 kg   = 2.20462 lb
1 oz   = 28.349 gm
1 ton   = 2240 lb = 1.01605 t (tonne) = 1016.05 kg
1 ton (US)   = 2000 lb = 907.185 kg

Force:
1 kgf   = 9.80665 N = 2.20462 lbf = 1 kp (kilopond)
1 lbf   = 4.44822 N
1 pdl (poundal) = 0.138255 N

Density:
1 kgf/m^3 = 0.062428 lb/ft^3
1 lb/ft^3 = 16.0185 kg/m^3
1 g/cm^3 = 1000 kg/m^3

Pressure, stress:
1 kgf/cm^2 = 98.0665 kPa (kN/m^2)
1 kgf/mm^2 = 9.80665 N/mm^2 = 1422.33 lbf/in^2 = 0.63497 tonf/in^2
1 lbf/in^2 (psi)   = 6.89476 kPa (kN/m^2)
1 Pa (N/m^2)   = 0.000145038 lbf/in^2
1 in w.g. (in H$_2$O) = 249.089 Pa
1 N/mm^2 = 1 MPa = 145.038 lbf/in^2 = 0.06475 tonf/in^2
                   = 0.10197 kgf/cm^2

Power:
1 kW   = 3412 Btu/hr
1 hp (horsepower) = 0.745700 kW

Energy, heat, work:
1 Btu   = 1.05506 kJ
1 cal   = 4.1868 J
1 kWh   = 3.6 MJ = 3412 Btu
1 therm = 100 000 Btu = 105.506 MJ
1 kJ   = 0.277778 W.h

Specific heat capacity, heat transfer:
1 cal/g°C   = 1 kcal/kg°C = 4186.8 J/kg.K
1 Btu/lb°F   = 4186.8 J/kg.K
1 Btu/h   = 0.293071 W
1 cal/cm.s°C   = 418.68 W/m.K (thermal conductivity)

1 Btu.in/ft^{2}h°F = 0.144228 W/m.K (thermal conductivity)
1 Btu/ft^{2}h°F   = 5.67826 W/m^2.K (heat transfer coeff.)

Miscellaneous:
1 std.atmos.   = 101.325 kPa = 760 mm Hg = 1.01325 bar
1 bar   = 100 kPa = 14.5038 lbf/in^2
1 cP (centipoise) = 1 mPa.s
1 cSt (centistoke) = 1 mm^2/s
1 cycle/s   = 1 Hz (Hertz)
1 hp   = 745.7 W

Useful approximations:

1 Btu	= 1 kJ	1 kg	= $2^1/_4$ lb
1 ft	= 30 cm	1 kgf	= 10 N
1 gal	= $4^1/_2$ l	1 std atmos.	= 1 bar
1 ha	= $2^1/_2$ acre	1 km	= $^5/_8$ mile
1 hp	= $^3/_4$ kW	1 litre	= $1^3/_4$ pint
1 in	= 25 mm	1 lbf	= $4^1/_2$ N
1 therm	= 100 MJ	1 yd	= 0.9 m
1 tonf/in^2	= 15 N/mm^2		
1 psi (lbf/in^2)	= 7 kPa		
1 N (newton)	= the weight of a small apple!		

Temperature:
°F   = 1.8 × °C + 32
°C   = (°F − 32)/1.8
0°C (Celsius) = 273.15 K (Kelvin)

Conversion table of stress values

Equivalent stresses

American (lb/in^2)	British (ton/in^2)	Metric (kgf/mm^2)	SI (N/mm^2)
250	0.112	0.176	1.724
500	0.223	0.352	3.447
1000	0.446	0.703	6.895
2000	0.893	1.406	13.789
3000	1.339	2.109	20.684
4000	1.788	2.812	27.579
5000	2.232	3.515	34.474
10 000	4.464	7.031	68.947
15 000	6.696	10.546	103.421
20 000	8.929	14.062	137.894
25 000	11.161	17.577	172.368
30 000	13.393	21.092	206.841
35 000	15.652	24.608	241.315
40 000	17.875	28.123	275.788
45 000	20.089	31.639	310.262
50 000	22.321	35.154	344.735
55 000	24.554	38.670	379.209
60 000	26.786	42.185	413.682
65 000	29.018	45.700	448.156
70 000	31.250	49.216	482.629
75 000	33.482	52.731	517.103
80 000	35.714	56.247	551.576
85 000	37.946	59.762	586.050
90 000	40.179	63.277	620.523
95 000	42.411	66.793	654.997
100 000	44.643	70.308	689.470

Conversions

10 000	4.464	7.031	68.947
22 399	**10**	15.749	154.438
14 223	6.349	**10**	98.066
14 504	6.475	10.197	**100**

Areas and volumes of circles, spheres, cylinders etc.

$\pi = 3.14159$ (approximation: $22/7$)

1 radian = 57.296 degrees

Circle; radius r, diameter d:

circumference $= 2\pi r = \pi d$

area $= \pi r^2 = \pi/4 \times d^2$

Sphere; radius r:

surface area $= 4\ \pi r^2$

volume $= {}^4/_3\ \pi r^3$

Cylinder; radius of base r, height h:

area of curved surface $= 2\pi rh$

volume $= \pi r^2 h$

Cone; radius of base r, height h:

volume $= {}^1/_2$ area of base $\times$ height

$= {}^1/_2\ \pi r^2 h$

Triangle; base b, height h:

area $= {}^1/_2 bh$

The physical properties of metals

Element	Symbol	Atomic weight	Melting point (°C)	Boiling point (°C)	Latent heat of fusion		Mean specific heat 0–100°C	
					(kJ/kg)	(cal/g)	(kJ/kg.K)	(cal/g°C)
Aluminium	Al	26.97	660.4	2520	386.8	92.4	0.917	0.219
Antimony	Sb	121.76	630.7	1590	101.7	24.3	0.209	0.050
Arsenic	As	74.93	volat.	616	–	–	0.331	0.079
Barium	Ba	137.37	729	2130	–	–	0.285	0.068
Beryllium	Be	9.02	1287	2470	133.5	31.9	2.052	0.490
Bismuth	Bi	209.0	217.4	1564	54.4	13.0	0.125	0.030
Cadmium	Cd	112.41	321.1	767	58.6	14.0	0.233	0.056
Calcium	Ca	40.08	839	1484	328.6	78.5	0.624	0.149
Carbon	C	12.01	–	–	–	–	0.703	0.168
Cerium	Ce	140.13	798	3430	–	–	0.188	0.045
Chromium	Cr	52.01	1860	2680	132.7	31.7	0.461	0.110
Cobalt	Co	58.94	1494	2930	244.5	58.4	0.427	0.102
Copper	Cu	63.57	1085	2560	180.0	43.0	0.386	0.092
Gallium	Ga	69.74	29.7	2205	80.2	19.2	0.377	0.090
Gold	Au	197.2	1064.4	2860	67.4	16.1	0.130	0.031
Indium	In	114.8	156	2070	–	–	0.243	0.058
Iridium	Ir	193.1	2447	4390	–	–	0.131	0.031
Iron	Fe	55.84	1536	2860	200.5	47.9	0.456	0.109
Lead	Pb	207.22	327.5	1750	20.9	5.0	0.130	0.031
Lithium	Li	6.94	181	1342	137.4	32.8	3.517	0.840
Magnesium	Mg	24.32	649	1090	194.7	46.5	1.038	0.248
Manganese	Mn	54.93	1244	2060	152.8	36.5	0.486	0.116
Mercury	Hg	200.61	−38.9	357	12.6	3.0	0.138	0.033
Molybdenum	Mo	96.0	2615	4610	–	–	0.251	0.060
Nickel	Ni	58.69	1455	2915	305.6	73.0	0.452	0.108
Niobium	Nb	92.91	2467	4740	–	–	0.268	0.064
Osmium	Os	190.9	3030	5000	–	–	0.130	0.031
Palladium	Pd	106.7	1554	2960	150.7	36.0	0.247	0.059
Phosphorus	P	31.04	44.1	279	20.9	5.0	0.791	0.189
Platinum	Pt	195.23	1770	3830	113.0	27.0	0.134	0.032
Potassium	K	39.1	63.2	759	67.0	16.0	0.754	0.180
Rhodium	Rh	102.91	1966	3700	–	–	0.243	0.058
Silicon	Si	28.3	1412	3270	502.4	120.0	0.729	0.174
Silver	Ag	107.88	961.9	2163	92.1	22.0	0.234	0.055
Sodium	Na	23.00	97.8	883	115.1	27.5	1.227	0.293
Strontium	Sr	87.63	770	1375	–	–	0.737	0.176
Sulphur	S	32.0	115	444.5	32.7	9.0	0.068	0.016
Tantalum	Ta	180.8	2980	5370	154.9	37.0	0.142	0.034
Tellurium	Te	127.6	450	988	31.0	7.4	0.134	0.032
Thallium	Tl	204	304	1473	–	–	0.130	0.031
Tin	Sn	118.7	232	2625	61.1	14.6	0.226	0.054
Titanium	Ti	47.9	1667	3285	376.8	90.0	0.528	0.126
Tungsten	W	184.0	3387	5555	167.5	40.0	0.138	0.033
Uranium	U	238.2	1132	4400	–	–	0.117	0.029
Vanadium	V	50.95	1902	3410	334.9	80.0	0.498	0.119
Zinc	Zn	65.38	419.6	911	110.1	26.3	0.394	0.094
Zirconium	Zr	90.6	1852	4400	–	–	0.289	0.069

The physical properties of metals (*continued*)

Element	Thermal conductivity (W/m.K)	Resistivity (µohm.cm at 20°C)	Vol. change on melting (%)	Density (g/cm³)	Coeff. of expansion (× 10⁻⁶/K)	Brinell hardness no.
Al	238	2.67	6.6	2.70	23.5	17
Sb	23.8	40.1	1.4	6.68	11	30
As	–	33.3	–	5.73	5.6	–
Ba	–	60	–	3.5	18	–
Be	194	3.3	–	1.85	12	–
Bi	9	117	–3.3	9.80	13.4	9
Cd	103	7.3	4.7	8.64	31	20
Ca	125	3.7	–	1.54	22	13
C	16.3	–	–	2.30	7.9	–
Ce	11.9	85.4	–	6.75	8	–
Cr	91.3	13.2	–	7.10	6.5	350
Co	96	6.3	–	8.90	12.5	125
Cu	397	1.69	4.1	8.96	17	48
Ga	41	–	–	5.91	18.3	–
Au	316	2.2	5.2	19.3	14.1	18.5
In	80	8.8	–	7.3	24.8	1
Ir	147	5.1	–	22.4	6.8	172
Fe	78	10.1	5.5	7.87	12.1	66
Pb	35	20.6	3.4	11.68	29	5.5
Li	76	9.3	1.5	0.53	56	–
Mg	156	4.2	4.2	1.74	26	25
Mn	7.8	160	–	7.4	23	–
Hg	8.7	96	3.75	13.55	61	–
Mo	137	5.7	–	10.2	5.1	147
Ni	89	6.9	–	8.9	13.3	80
Nb	54	16	–	8.6	7.2	–
Os	87	8.8	–	22.5	4.6	–
Pd	75	10.8	–	12.0	11.0	50
P	–	–	–	1.83	6.2	–
Pt	73	10.6	–	21.45	9.0	52
K	104	6.8	2.8	0.86	83	0.04
Rh	148	4.7	–	12.4	8.5	156
Si	139	10³–10⁶	–	2.34	7.6	–
Ag	425	1.6	4.5	10.5	19.1	25
Na	128	4.7	2.5	0.97	71	0.1
Sr	–	23	–	2.6	100	–
S	272	–	–	2.07	70	–
Ta	58	13.5	–	16.6	6.5	40
Te	3.8	1.6×10^5	–	6.24	–	–
Tl	45.5	16.6	–	11.85	30	–
Sn	73.2	12.6	2.8	7.3	23.5	–
Ti	21.6	54	–	4.5	8.9	–
W	174	5.4	–	19.3	4.5	–
U	28	27	–	19.0	–	–
V	31.6	19.6	–	6.1	8.3	–
Zn	120	6.0	6.5	7.14	31	35
Zr	22.6	44	–	6.49	5.9	–

Densities of casting alloys

Alloy	BS1490	g/ml	Alloy	BS1400	g/ml
Aluminium alloys			*Copper alloys*		
Pure Al		2.70	HC copper	HCC1	8.9
Al–Si5Cu3	LM4	2.75	Brass CuZn38Al	DCB1	8.5
Al–Si7Mg	LM25	2.68	CuZn33Pb2Si	HTB1	8.5
Al–Si8Cu3Fe	LM24	2.79	CuZn33Pb2	SCB3	8.5
AlSi12	LM6	2.65	Phosphor bronze		
			CuSn11P	PB1	8.8
Cast steels			CuSn12	PB2	8.7
Low carbon <0.20		7.86	Lead bronze		
Med. carbon 0.40		7.86	CuSn5Pb20	LB5	9.3
High carbon >0.40		7.84	Al bronze		
			CuAl10Fe2	AB1	7.5
Low alloy		7.86	Gunmetal		
Med. alloy		7.78	CuSnPb5Zn5	LG2	8.8
Med./high alloy		7.67	Copper nickel		
			CuNi30Cr2FeMnSi	CN1	8.8
Stainless					
13Cr		7.61	*Cast irons*		
18Cr8Ni		7.75	Grey iron 150 MPa		6.8–7.1
			200		7.0–7.2
Other alloys			250		7.2–7.4
Zinc base			300		7.3–7.4
ZnAl4Cu1		6.70	Whiteheart malleable		7.45
			Blackheart malleable		7.27
Lead base			White iron		7.70
PbSb6		10.88	Ductile iron (s.g.)		7.2–7.3
Tin base (Babbit)		7.34	Ni-hard		7.6–7.7
Inconel Ni76Cr18		8.50	High silicon (15%)		6.8

Approximate bulk densities of common materials

Material	kg/m^3	lb/ft^3	Material	kg/m^3	lb/ft^3
Aluminium, cast	2560	160	Lead	11370	710
wrought	2675	167	Limestone	2530–2700	158–168
Aluminium bronze	7610	475			
Ashes	590	37	Magnesite	2530	158
			Mercury	13 560	847
Brass, rolled	8390	524	Monel	8870	554
swarf	2500	175			
Babbit metal	7270	454	Nickel, cast	8270	516
Brick, common	1360–1890	85–118	Nickel silver	8270	516
fireclay	1840	115			
Bronze	8550	534	Phosphor bronze	8580	536
			Pig iron, mean	4800	300
Cast iron, solid	7210	450	Pig iron and scrap		
turnings	2240	140	(cupola charge)	5400	336
Cement, loose	1360	85			
Chalk	2240	140	Sand, moulding	1200–1440	75–90
Charcoal, lump	290	18	silica	1360–1440	85–90
Clay	1900–2200	120–135	Silver, cast	10 500	656
Coal	960–1280	60–80	Steel	7850	490
Coal dust	850	53			
Coke	450	28	Tin	7260	453
Concrete	2240	140			
Copper, cast	8780	548	Water, ice	940	58.7
Cupola slag	2400	150	liquid 0°C	1000	62.4
			100°C	955	59.6
Dolomite	2680	167	Wood, balsa	100–130	7–8
			oak	830	52
Fire clay	1440	90	pine	480	30
French chalk	2600	162	teak	640	40
			Wrought iron	7700	480
Glass	2230	139			
Gold, pure	19 200	1200	Zinc, cast	6860	428
22 carat	17 500	1090	rolled	7180	448
Graphite, powder	480	30			
solid	2200	138			

Patternmakers' contraction allowances

Castings are always smaller in dimensions than the pattern from which they are made, because as the metal cools from its solidification temperature to room temperature, thermal contraction occurs. Patternmakers allow for this contraction by making patterns larger in dimensions than the required castings by an amount known as the "contraction allowance". Originally this was done by making use of specially engraved rules, known as "contraction rules", the dimensions of which incorporated a contraction allowance such as 1 in 75 for aluminium alloys, or 1 in 96 for iron castings. Nowadays, most patterns and coreboxes are made using computer-controlled machine tools and it is more convenient to express the contraction as a percentage allowance.

Predicting casting contraction can never be precise, since many factors are involved in determining the exact amount of contraction that occurs. For example, when iron castings are made in greensand moulds, the mould walls may move under the pressure of the liquid metal, causing expansion of the mould cavity, thus compensating for some of the metal contraction. Cored castings may not contract as much as expected, because the presence of a strong core may restrict movement of the casting as it is cooling. Some core binders expand with the heat of the cast metal causing the casting to be larger than otherwise expected. For these reasons, and others, it is only possible to predict contractions approximately, but if a patternmaker works with a particular foundry for a long period, he will gain experience with the foundry's design of castings and with the casting methods used in the foundry. Based on such experience, more precise contraction allowances can be built into the patterns.

The usually accepted contraction allowances for different alloys are given in the following table.

Alloy		Contraction allowance (%)
Aluminium alloys		
Al–Si5Cu3	LM4	
Al–Si7Mg	LM25	1.3
Al–Si8Cu3Fe	LM24	
Al–Si12	LM6	
Beryllium copper		1.6
Bismuth		1.3
Brass		1.56
Bronze, aluminium		2.32
manganese		0.83–1.56
phosphor		1.0–1.6
silicon		1.3–1.6
Cast iron, grey		0.9–1.04
white		2.0
ductile (s.g.)		0.6–0.8
malleable		1.0–1.4
Copper		1.6
Gunmetal		1.0–1.6
Lead		2.6
Magnesium alloys		1.30–1.43
Monel		2.0
Nickel alloys		2.0
Steel, carbon		1.6–2.0
chromium		2.0
manganese		1.6–2.6
Tin		2.0
White metal		0.6
Zinc alloys		1.18

Volume shrinkage of principal casting alloys

Most alloys shrink in volume when they solidify, the shrinkage can cause voids in castings unless steps are taken to "feed" the shrinkage by the use of feeders.

Casting alloy	Volume shrinkage (%)
Carbon steel	6.0
Alloyed steel	9.0
High alloys steel	10.0
Malleable iron	5.0
Al	8.0
Al–Cu4Ni2Mg	5.3
Al–Si12	3.5
Al–Si5Cu2Mg	4.2
Al–Si9Mg	3.4
Al–Si5Cu1	4.9
Al–Si5Cu2	5.2
Al–Cu4	8.8
Al–Si10	5.0
Al–Si7NiMg	4.5
Al–Mg5Si	6.7
Al–Si7Cu2Mg	6.5
Al–Cu5	6.0
Al–Mg1Si	4.7
Al–Zn5Mg	4.7
Cu (pure)	4.0
Brass	6.5
Bronze	7.5
Al bronze	4.0
Sn bronze	4.5

Comparison of sieve sizes

Sieves used for sand grading are of 200 mm diameter and are now usually metric sizes, designated by their aperture size in micrometres (μm). The table lists sieve sizes in the British Standard Metric series (BS410 : 1976) together with other sieve types.

Sieve aperture, micrometres and sieve numbers

ISO/R.565 series (BS410:1976) (μm)	BSS		ASTM	
	No.	μm	No.	μm
(1000)	16	1003	18	1000
710	22	699	22	710
500	30	500	30	500
355	44	353	45	350
250	60	251	60	250
(212)	72	211	70	210
180				
(150)	100	152	100	149
125			120	125
90	150	104	150	105
63	200	76	200	74
(45)	300	53	325	44

Notes: The 1000 and 45 sieves are optional.
The 212 and 150 sieves are also optional, but may be included to give better separation between the 250 and 125 sieves.

Calculation of average grain size

The adoption of the ISO metric sieves means that the old AFS grain fineness number can no longer be calculated. Instead, the average grain size, expressed as micrometres (μm) is now used. This is determined as follows:

1 Weigh a 100 g sample of dry sand.
2 Place the sample into the top sieve of a nest of ISO sieves on a vibrator. Vibrate for 15 minutes.
3 Remove the sieves and, beginning with the top sieve, weigh the quantity of sand remaining on each sieve.
4 Calculate the percentage of the sample weight retained on each sieve, and arrange in a column as shown in the example.
5 Multiply the percentage retained by the appropriate multiplier and add the products.
6 Divide by the total of the percentages retained to give the average grain size.

Example

ISO aperture (μm)	Percentage retained	Multiplier	Product
≥ 710	trace	1180	0
500	0.3	600	180
355	1.9	425	808
250	17.2	300	5160
212	25.3	212	5364
180	16.7	212	3540
150	19.2	150	2880
125	10.6	150	1590
90	6.5	106	689
63	1.4	75	105
≤63	0.5	38	19
Total	99.6	–	20 335

Average grain size = 20 335/99.6
            = 204 μm

Calculation of AFS grain fineness number

Using either the old BS sieves or AFS sieves, follow, steps 1–4 above.
5 Arrange the results as shown in the example below.
6 Multiply each percentage weight by the preceding sieve mesh number.
7 Divide by the total of the percentages to give the AFS grain fineness number.

Example

BS sieve number	% sand retained on sieve	Multiplied by previous sieve no.	Product
10	nil	–	–
16	nil	–	–
22	0.2	16	3.2
30	0.8	22	17.6
44	6.7	30	201.0
60	22.6	44	1104.4
100	48.3	60	2898.0
150	15.6	100	1560.0
200	1.8	150	270.0
pan	4.0	200	800.0
Total	100.0	–	6854.2

AFS grain fineness number = 6854.2/100

$$= 68.5 \text{ or } 68 \text{ AFS}$$

Foundry sands usually fall into the range 150–400 µm, with 220–250 µm being the most commonly used. Direct conversion between average grain size and AFS grain fineness number is not possible, but an approximate relation is shown below:

AFS grain fineness no.	35	40	45	50	55	60	65	70	80	90
Average grain size (µm)	390	340	300	280	240	220	210	195	170	150

While average grain size and AFS grain fineness number are useful parameters, choice of sand should be based on particle size distribution.

Recommended standard colours for patterns

Part of pattern		*Colour*
As-cast surfaces which are to be left unmachined		Red or orange
Surfaces which are to be machined		Yellow
Core prints for unmachined openings and end prints		
	Periphery	Black
	Ends	Black
Core prints for machined openings	Periphery	Yellow stripes on black
	Ends	Black
Pattern joint (split patterns)	Cored section	Black
	Metal section	Clear varnish
Touch core	Cored shape	Black
	Legend	"Touch"
Seats of and for loose pieces and loose core prints		Green
Stop offs		Diagonal black stripes with clear varnish
Chilled surfaces	Outlined in	Black
	Legend	"Chill"

Dust control in foundries

Air extraction is used in foundries to remove silica dust from areas occupied by operators. The following table indicates the approximate air velocities needed to entrain sand particles.

Terminal velocities of spherical particles of density 2.5 g/cm^3 (approx.)

BS sieve size	Particle dia. (μm)	Terminal velocity m/sec	ft/sec	ft/min
16	1003	7.0	23	1400
30	500	4.0	13	780
44	353	2.7	9	540
60	251	1.8	6	360
100	152	1.1	3.5	210
150	104	0.5	1.7	100
200	76	0.4	1.3	80

For the comfort and safety of operators, air flows of around 0.5 m/sec are needed to carry away silica dust. If air flow rate is too high, around the shake-out for example, there is a danger that the grading of the returned sand will be altered.

Buoyancy forces on cores

When liquid metal fills a mould containing sand cores, the cores tend to float and must be held in position by the core prints or by chaplets. The following table lists the buoyancy forces experienced by silica sand cores in various liquid metals, expressed as a proportion of the weight of the core:

Liquid metal	Ratio of buoyant force to core weight
Aluminium	0.66
Brass	4.25
Copper	4.50
Cast iron	3.50
Steel	3.90

Core print support

Moulding sand (green sand) in a core print will support about 150 kN/m^2 (21 psi). So the core print can support the following load:

Support (kN) = Core print area (m^2) × 150

1 kN = 100 kgf (approx.)

Support (kgf) = Core print area (m^2) × 15 000

Example: A core weighing 50 kg has a core print area of 10 × 10 cm (the area of the upper, support surface), i.e. 0.1 × 0.1 = 0.01 m^2. The print support is 150 × 0.01 = 1.5 kN = 150 kgf.

If the mould is cast in iron, the buoyancy force is 50 × 3.5 = 175 kgf so chaplets may be necessary to support the core unless the print area can be increased.

Opening forces on moulds

Unless a mould is adequately clamped or weighted, the force exerted by the molten metal will open the boxes and cause run-outs. If there are insufficient box bars in a cope box, this same force can cause other problems like distortion and sand lift. It is important therefore to be able to calculate the opening force so that correct weighting or clamping systems can be used.

The major force lifting the cope of the mould is due to the metallostatic pressure of the molten metal. This pressure is due to the height, or head, of metal in the sprue above the top of the mould (H in Fig. 1.1). Additional forces exist from the momentum of the metal as it fills the mould and from forces transmitted to the cope via the core prints as the cores in cored castings try to float.

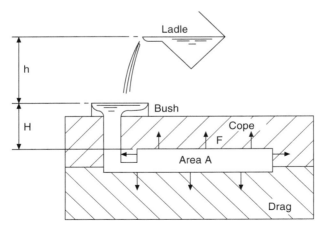

Figure 1.1   *Opening forces of moulds.*

The momentum force is difficult to calculate but can be taken into account by adding a 50% safety factor to the metallostatic force.

The opening metallostatic force is calculated from the total upward-facing area of the cope mould in contact with the metal (this includes the area of all the mould cavities in the box). The force is:

$$F(\text{kgf}) = \frac{A \times H \times d \times 1.5}{1000}$$

A is the upward facing area in cm^2
H (cm) is the height of the top of the sprue above the average height of the upward facing surface
d is the density of the molten metal (g/cm^3)
1.5 is the "safety factor"

For ferrous metals, d is about 7.5, so:

$$F(\text{kgf}) = \frac{11 \times A \times H}{1000}$$

For aluminium alloys, d is about 2.7, so:

$$F(\text{kgf}) = \frac{4 \times A \times H}{1000}$$

Forces on cores

The core tends to float in the liquid metal and exerts a further upward force (see page 18)
    In the case of ferrous castings, this force is

$$3.5 \times W \ (\text{kgf})$$

where W is the weight of the cores in the mould (in kg).
    In aluminium, the floating force can be neglected.
    The total resultant force on the cope is (for ferrous metals)

$$(11 \times A \times H)/1000 + 3.5 \ W \ \text{kgf}$$

Example: Consider a furane resin mould for a large steel valve body casting having an upward facing area of 2500 cm^2 and a sprue height (H) of 30 cm with a core weighing 40 kg. The opening force is

$$11 \times 2500 \times 30/1000 + 3.5 \times 40 = 825 + 140$$

$$= 965 \ \text{kgf}$$

It is easy to see why such large weights are needed to support moulds in jobbing foundries.

Dimensional tolerances and consistency achieved in castings

Errors in dimensions of castings are of two kinds:

Accuracy: the variation of the mean dimension of the casting from the design dimension given on the drawing

Consistency: statistical errors, comprising the dimensional variability round the mean dimension

Dimensional accuracy

The major causes of deviations of the mean dimension from the target value are contraction uncertainty and errors in pattern dimensions. It is usually possible to improve accuracy considerably by alternations to pattern equipment after the first sample castings have been made.

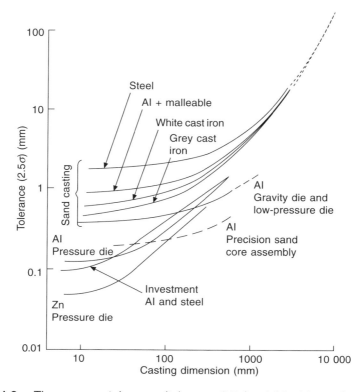

Figure 1.2 *The average tolerance (taken as 2.5σ) exhibited by various casting processes. (From Campbell, J. (1991) Castings, Butterworth-Heinemann, reproduced by permission of the publishers.)*

Dimensional consistency

Changes in process variables during casting give rise to a statistical spread of measurements about a mean value. If the mean can be made to coincide with the nominal dimension by pattern modification, the characteristics of this statistical distribution determine the tolerances feasible during a production run.

The consistency of casting dimensions is dependent on the casting process used and the degree of process control achieved in the foundry. Figure 1.2 illustrates the average tolerance exhibited by various casting processes. The tolerance is expressed as 2.5σ (2.5 standard deviations), meaning that only 1 casting in 80 can be expected to have dimensions outside the tolerance.

There is an International Standard, ISO 8062–1984(E) *Castings – System of dimensional tolerances,* which is applicable to the dimensions of cast metals and their alloys produced by sand moulding, gravity diecasting, low pressure diecasting, high pressure diecasting and investment casting. The Standard defines 16 tolerance grades, designated CT1 to CT16, listing the total casting tolerance for each grade on raw casting dimensions from 10 to 10 000 mm. The Standard also indicates the tolerance grades which can be expected for both long and short series production castings made by various processes from investment casting to hand-moulded sand cast.

Reference should be made to ISO 8062 or the equivalent British Standard BS6615:1985 for details.

Chapter 2
Types of cast iron

Introduction

The first iron castings to be made were cast directly from the blast furnace. Liquid iron from a blast furnace contains around 4%C and up to 2%Si, together with other chemical elements derived from the ore and other constituents of the furnace charge. The presence of so much dissolved carbon etc. lowers the melt point of the iron from 1536°C (pure iron) to a eutectic temperature of about 1150°C (Fig. 2.1) so that blast furnace iron is fully liquid and highly fluid at temperatures around 1200°C. When the iron solidifies, most of the carbon is thrown out of solution in the form either of graphite or of iron carbide, Fe_3C, depending on the composition of the iron, the rate of cooling from liquid to solid and the presence of nucleants.

If the carbon is precipitated as flake graphite, the casting is called 'grey iron', because the fractured surface has a dull grey appearance due to the presence of about 12% by volume of graphite. If the carbon precipitates as carbide, the casting is said to be 'white iron' because the fracture has a shiny white appearance. In the early days of cast iron technology, white iron was of little value, being extremely brittle and so hard that it was unmachinable. Grey iron, on the other hand, was soft and readily machined and although it had little ductility, it was less brittle than white iron.

Iron castings were made as long ago as 500 BC (in China) and from the 15th century in Europe, when the blast furnace was developed. The great merits of grey iron as a casting alloy, which still remain true today, are its low cost, its high fluidity at modest temperatures and the fact that it freezes with little volume change, since the volume expansion of the carbon precipitating as graphite compensates for the shrinkage of the liquid iron. This means that complex shapes can be cast without shrinkage defects. These factors, together with its free-machining properties, account for the continuing popularity of grey cast iron, which dominates world tonnages of casting production (Table 2.1).

Greater understanding of the effect of chemical composition and of nucleation of suitable forms of graphite through inoculation of liquid iron, has vastly improved the reliability of grey iron as an engineering material. Even so, the inherent lack of ductility due to the presence of so much graphite precipitated in flake form (Fig. 2.2) limits the applications to which grey iron can be put.

A malleable, or ductile form of cast iron was first made by casting 'white

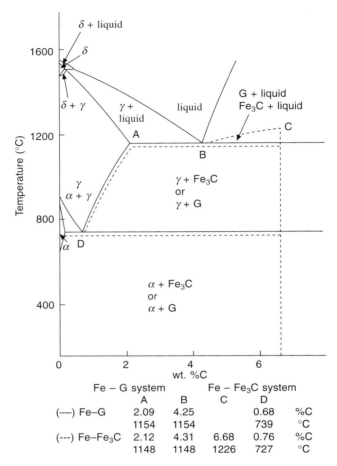

Figure 2.1   *The iron–carbon phase diagram. (From Elliott, R.,* Cast Iron Technology, *1988, Butterworth-Heinemann, reproduced by permission of the publishers.)*

Table 2.1   Breakdown of iron casting tonnages 1996 (1000s tonnes)

	Total iron	*Grey iron*	*Ductile iron*	*Malleable iron*
Germany France UK	6127	3669 (59.9%)	2368 (38.6%)	84 (1.37%)
USA	10 314	6048 (58.6%)	4034 (39.1%)	232 (2.25%)

Data from CAEF report *The European Foundry Industry 1996*
US data from Modern Castings

iron' and then by a long heat treatment process, converting the iron carbide to graphite. Under the right conditions the graphite developed in discrete, roughly spherical aggregates (Fig. 2.3) so that the casting became ductile with elongation of 10% or more. The first malleable iron, 'whiteheart iron'

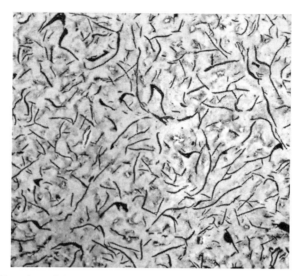

Figure 2.2 *Random flake graphite, 4% picral, ×100. (From BCIRA Broadsheet 138, reproduced by courtesy of CDC.)*

was made by Réaumur in France in 1720. The more useful 'blackheart' malleable iron was developed in the USA by Boyden around 1830. Malleable cast iron became a widely used casting alloy wherever resistance to shock loading was required. It was particularly suitable for transmission components for railways and automotive applications.

A major new development occurred in the late 1940s with the discovery that iron having a nodular form of graphite could be cast directly from the

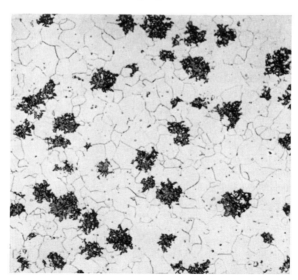

Figure 2.3 *Malleable cast iron, 4% picral, ×100. (From BCIRA Broadsheet 138, reproduced by courtesy of CDC.)*

melt after treatment of liquid iron of suitable composition with magnesium. (Fig. 2.4). The use of 'spheroidal graphite' or 'nodular' iron castings has since grown rapidly as the technology became understood and 'ductile iron', as it is now generally known, has gained a large and still growing, sector of total cast iron production (Table 2.1).

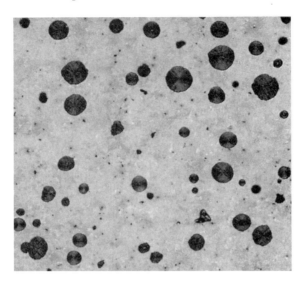

Figure 2.4   *Nodular graphite, 4% picral ×100. (From BCIRA Broadsheet 138, reproduced by courtesy of CDC.)*

The great hardness and abrasion resistance of white iron has also been exploited. The strength of white iron has been improved through alloying and heat treatment, and white iron castings are widely used in applications such as mineral processing, shot blasting etc. where the excellent wear resistance can be fully used.

Finally there are a number of special cast irons designed to have properties of heat resistance, or acid resistance etc. In the following chapters, each type of iron will be considered separately and its method of production described.

The mechanical properties of cast iron are derived mainly from the matrix and irons are frequently described in terms of their matrix structure, that is, ferritic or pearlitic:

Ferrite is a Fe–C solid solution in which some Si, Mn, Cu etc. may also dissolve. It is soft and has relatively low strength. Ferritic irons can be produced as-cast or by annealing.

Pearlite is a mixture of lamellae of ferrite and Fe_3C formed from austenite by a eutectoid reaction. It is relatively hard and the mechanical properties of a pearlitic iron are affected by the spacing of the pearlite lamellae, which is affected by the rate of cooling of the iron from the eutectoid temperature of around 730°C.

Ferrite–Pearlite mixed structures are often present in iron castings. *Bainite* is usually formed by an austempering heat treatment (normally on spheroidal graphite irons) and produces high tensile strength with toughness and good fatigue resistance.

Austenite is retained when iron of high alloy (nickel and chromium) content cools. Heat and corrosion resistance are characteristics of austenitic irons.

Physical properties of cast irons

The physical properties of cast irons are affected by the amount and form of the graphite and the microstructure of the matrix. Tables 2.2, 2.3, 2.4, 2.5 and 2.6 show, respectively, the density, electrical resistivity, thermal expansion, specific heat capacity and thermal conductivity of cast irons. The figures in the tables should be regarded as approximate.

Table 2.2 Density of cast irons

	Grey iron						
Tensile strength (N/mm^2)	150	180	220	260	300	350	400
Density at 20°C (g/cm^3)	7.05	7.10	7.15	7.20	7.25	7.30	7.30

	Ductile iron				
Grade	350/22	400/12	500/7	600/3	700/2
Density at 20°C (g/cm^3)	7.10	7.10	7.10–7.17	7.17–7.20	7.20

	Malleable iron				
Grade	350/10	450/6	550/4	600/3	700/2
Density at 20°C (g/cm^3)	7.35	7.30	7.30	7.30	7.30

	Other cast irons				
Type	White cast irons			Austenitic (Ni-hard)	Grey high-Si (6%)
	Unalloyed	15–30%Cr	Ni-Cr		
Density at 20°C (g/cm^3)	7.6–7.8	7.3–7.5	7.6–7.8	7.4–7.6	6.9–7.2

Table 2.3 Electrical resistivity of cast irons

	Grey iron						
Tensile strength (N/mm^2)	150	180	220	260	300	350	400
Resistivity at 20°C (micro-ohms.m^2/m)	0.80	0.78	0.76	0.73	0.70	0.67	0.64

	Ductile iron				
Grade	350/22	400/12	500/7	600/3	700/2
Resistivity at 20°C (micro-ohms.m^2/m)	0.50	0.50	0.51	0.53	0.54

	Malleable iron				
Grade	350/10	450/6	550/4	600/3	700/2
Resistivity at 20°C (micro-ohms.m^2/m)	0.37	0.40	0.40	0.41	0.41

Table 2.4 Coefficient of linear thermal expansion for cast irons

Type of iron	Typical coefficient of linear expansion for temperature ranges (10^{-6} per °C)				
	20–100°C	20–200°C	20–300°C	20–400°C	20-500°C
Ferritic flake or nodular	11.2	11.9	12.5	13.0	13.4
Pearlitic flake or nodular	11.1	11.7	12.3	12.8	13.2
Ferritic malleable	12.0	12.5	12.9	13.3	13.7
Pearlitic malleable	11.7	12.2	12.7	13.1	13.5
White iron	8.1	9.5	10.6	11.6	12.5
14–22% Ni austenitic	16.1	17.3	18.3	19.1	19.6
36% Ni austenitic	4.7	7.0	9.2	10.9	12.1

Table 2.5 Specific heat capacity of cast irons

Typical mean values for grey, nodular and malleable irons, from room temperature to 1000°C

Mean value for each temperature range (J/kg.K)

20–100°C	20–200°C	20–300°C	20–400°C	20–500°C	20–600°C	20–700°C	20–800°C	20–900°C	20–1000°C
515	530	550	570	595	625	655	695	705	720

Typical mean values for grey, nodular and malleable irons, for 100°C ranges

Mean value for each temperature range, (J/kg.K)

100–200°C	200–300°C	300–400°C	400–500°C	500–600°C	600–700°C	700–800°C	800–900°C	900–1000°C
540	585	635	690	765	820	995	750	850

Iron casting processes

The majority of repetition iron castings are made in green sand moulds with resin-bonded cores. The Croning resin shell moulding process is used where

Table 2.6   Thermal conductivity of cast irons

	Grey iron						
Tensile strength (N/mm²)	150	180	220	260	300	350	400
Thermal conductivity (W/m.K)							
100°C	65.6	59.5	53.6	50.2	47.7	45.3	45.3
500°C	40.9	40.0	38.9	38.0	37.4	36.7	36.0

	Ductile iron				
Grade	350/22	400/12	500/7	600/3	700/2
Thermal conductivity (W/m.K)					
100°C	40.2	38.5	36.0	32.9	29.8
500°C	36.0	35.0	33.5	31.6	29.8

	Malleable iron				
Grade	350/10	450/6	550/4	600/3	700/2
Thermal conductivity (W/m.K)					
100°C	40.4	38.1	35.2	34.3	30.8
500°C	34.6	34.1	32.0	31.4	28.9

high precision and good surface finish are needed. The Lost Foam Process is also used for repetition castings. Castings made in smaller numbers are made in chemically bonded sand moulds.

Special sand processes such as Vacuum Moulding and Full-Mould are used for certain iron castings and there are a few permanent mould (diecasting) foundries making iron castings, but the short die-life of only a few thousand components has restricted the use of ferrous diecasting.

Chapter 3
Grey cast iron

Specifications

The properties of grey iron castings are affected by their chemical composition and the cooling rate of the casting, which is influenced by the section thickness and shape of the casting. With the widespread acceptance of SI units, there has been a convergence of national specifications for grey cast iron based on minimum tensile strength measured in N/mm^2 (MPa) on a test piece machined from a separately cast, 30 mm diameter bar, corresponding to a relevant wall thickness of 15 mm. There is no requirement in terms of composition and the foundryman is free to make his own choice based on the requirements of the particular casting.

In 1997 a European Standard EN 1561:1997 was approved by CEN (European Committee for Standardization). This standard has been given the status of a national standard in countries which have CEN members. CEN members are the national standards bodies of Austria, Belgium, Czech Republic, Denmark, Finland, France, Germany, Greece, Iceland, Ireland, Italy, Luxembourg, Netherlands, Norway, Portugal, Spain, Sweden, Switzerland and United Kingdom.

In the UK, for example, the previously used standard BS 1452:1990 has been withdrawn and replaced by BS EN 1561:1997.

The six ISO grades are generally recognised in most countries. The USA continues to use grades based on tensile strengths measured in 1000s psi but now has corresponding metric standards. The data in Table 3.1 is intended to help foundrymen who are required to supply castings to other countries, however, it must be understood that while there are similarities between the various national standards, the grades are not necessarily identical. It is advisable to consult the original standards for details of methods of testing etc.

Many specifications also define grades of grey iron specified by their hardness, related to section thickness. In the European Standard EN 1561:97 Brinell Hardness HB 30 is used, the mandatory values apply to Relevant Wall Thickness of 40–80 mm (Table 3.2).

An approximate guide to variations in tensile strength with section thickness is given in Fig. 3.1.

Table 3.1 National standards

Country	Specification	Designation	Minimum tensile strength (N/mm^2)								
			100	150	180	200	250	275	300	350	400
Europe	EN 1561:1997	EN-GJL-	100	150		200	250		300	350	400
Japan	JIS G5501 1995	FC	100	150		200	250		300	350	
		Class	1	2		3	4		5	6	
Russia	GOST 1412 1979	Sch	10	15	18	20	25		30	35	40
USA	ASTM A48-94a	Grade		20	25	30	35	40	45	50	60
	ASTM A48M-94	Grade		150	175	200	225 250	275	300	325 350	400
International	ISO 185-1988	Grade	100	150		200	250		300	350	
Equivalent tonf/in^2			6.5	9.7		12.9	16.2		19.4	22.7	

Note: Consult National Specifications for full details.

Table 3.2   HB30 hardness grades of grey iron (EN 1561:1997)

Grade EN-GJL-		HB155	HB175	HB195	HB215	HB235	HB255
HB30	min	–	100	120	145	165	185
	max	155	175	195	215	235	255

Note: Consult EN 1561:1997 for details.

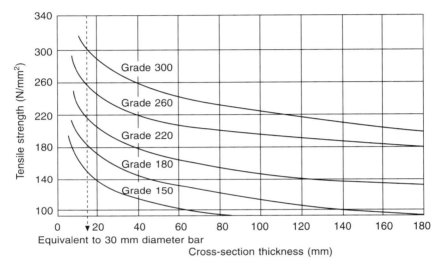

Figure 3.1   *Variation of tensile strength with section thickness for several grades of iron. (Data supplied by CDC.)*

Relationship between composition, strength and structure of grey cast iron

The properties of grey irons depend on the size, amount and distribution of the graphite flakes and on the structure of the metal matrix. These, in turn, depend on the chemical composition of the iron, in particular its carbon and silicon content; and also on processing variables such as method of melting, inoculation practice and the cooling rate of the casting.

Carbon equivalent

The three constituents of cast iron which most affect strength and hardness are total carbon, silicon and phosphorus. An index known as the 'carbon equivalent value' (CEV) combines the effects of these elements. The grey iron eutectic occurs at a carbon content of 4.3% in the binary Fe–C system. If silicon and phosphorus are present, the carbon content of the eutectic is lowered. The effect of Si and P contents on the carbon content of the eutectic are given by the expression

$$\text{carbon equivalent value (CEV)} = \text{T.C.\%} + \frac{\text{Si\%} + \text{P\%}}{3}$$

Cast irons with carbon equivalent values of less than 4.3% are called hypo-eutectic irons, those having carbon equivalent values higher than 4.3% are called hyper-eutectic irons. Since the structure (and hence the strength) of flake irons is a function of composition, a knowledge of the CEV of an iron can give an approximate indication of the strength to be expected in any sound section. This is conveniently expressed in graphical form (Fig. 3.2).

From Fig. 3.2 it is possible to construct a series of curves showing the reduction of strength with increasing section thickness, this is shown in Fig. 3.1 for the common grades of grey iron. It is important to realise that each of these curves represents an average figure and a band of uncertainty exists on either side. Nevertheless, these figures represent those likely to be achieved on test bars cast in green sand moulds.

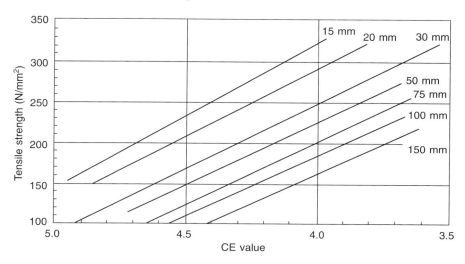

Figure 3.2 *Relationship between tensile strength and carbon equivalent value for various bar diameters. (Data supplied by CDC.)*

Note that the 'carbon equivalent liquidus value' (CEL) is frequently used as a shop floor method of quality control since it can be directly and readily measured. When unalloyed molten irons cool, a temperature arrest occurs when solidification commences, the liquidus arrest. The temperature of this arrest is related to the C, Si and P content of the iron by the expression

$$\text{carbon equivalent liquidus (CEL)} = \text{T.C.\%} + \text{Si\%}/4 + \text{P\%}/2$$

The CEL value is a guide to the tensile strength and chilling characteristics of the iron.

Machinability

In general, the higher the hardness, the poorer the machinability and castings with hardnesses above 250–260 are usually regarded as unsatisfactory. It is necessary to avoid mottled, chilled or white irons which contain free carbide, making them hard and unmachinable. Most grey iron castings are required to be strong and readily machinable, this is achieved with a pearlitic structure having no free ferrite or free carbide. Soft irons, freely machinable containing appreciable amounts of free ferrite as well as pearlite, are suitable for certain applications, particularly for heavy section castings (Fig. 3.3). (Figures 3.1, 3.2, 3.3 are reproduced by kind permission of CDC, Alvechurch, Birmingham.)

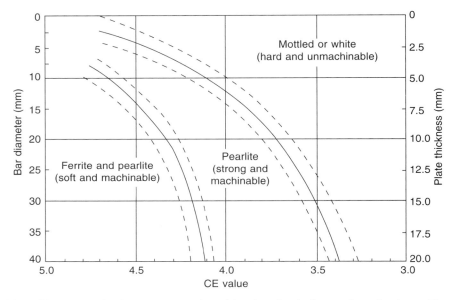

Note: The general microstructure of a wide plate is similar to that of a bar with a diameter equal to twice the thickness of the plate. At a surface however the chilling tendency of a plate casting will be greater and the structure can then be assumed to be similar to that of a bar with a diameter equal to the thickness of the plate. The left-hand scale is therefore used to assess surface chilling tendency and the right-hand scale for assessing the general structure for a plate casting of a given thickness.

Figure 3.3   *Relationship between section size, CEV, and structure. (Data supplied by CDC.)*

Classification of graphite flake size and shape

The properties of grey iron castings are influenced by the shape and distribution of the graphite flakes. The standard method of defining graphite forms is based on the system proposed by the American Society for the Testing of Metals, ASTM Specification A247 which classifies the form, distribution and size of the graphite. Certain requirements must be met before a sample is evaluated. Attention must be paid to the location of the microspecimen in relation to the rest of the casting, to the wall thickness and the distance from the as-cast surface. Care is also needed in grinding and polishing so that as much graphite as possible is retained in a representative cross-section. The specimen is normally examined in the polished, unetched condition at a magnification of 100× and with a field of view of about 80 mm.

Flake morphologies are divided into five classes (Fig. 3.4a).

Type A is a random distribution of flakes of uniform size, and is the preferred type for engineering applications. This type of graphite structure forms when a high degree of nucleation exists in the liquid iron, promoting solidification close to the equilibrium graphite eutectic.

Type B graphite forms in a rosette pattern. The eutectic cell size is large because of the low degree of nucleation. Fine flakes form at the centre of the rosette because of undercooling, these coarsen as the structure grows.

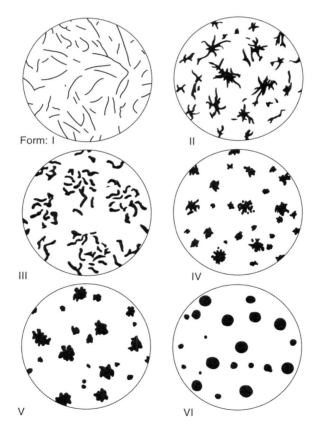

Figure 3.4 (a) *Reference diagrams for the graphite form (Distribution A). The diagrams show only the outlines and not the structure of the graphite.*

Dimensions of the graphite particles forms I to VI

Reference number	Dimensions of the particles observed at ×100 (mm)	True dimensions (mm)
1	>100	>1
2	50–100	0.5–1
3	25–50	0.25–0.5
4	12–25	0.12–0.25
5	6–12	0.06–0.12
6	3–6	0.03–0.06
7	1.5–3	0.015–0.03
8	<1.5	<0.015

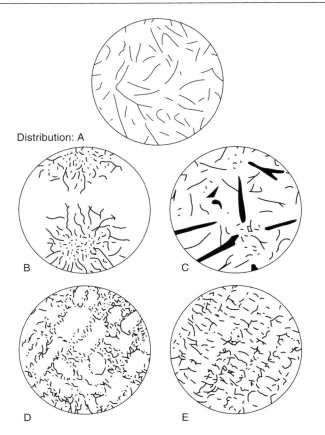

Figure 3.4 (b) *Reference diagrams for the distribution of graphite (Form 1). The diagrams show only the outlines and not the structure of the graphite.*

Type C structures occur in hyper-eutectic irons, where the first graphite to form is primary kish graphite. It may reduce tensile properties and cause pitting on machined surfaces.

Type D and Type E are fine, undercooled graphites which form in rapidly cooled irons having insufficient graphite nuclei. Although the fine flakes increase the strength of the eutectic, this morphology is undesirable because it prevents the formation of a fully pearlitic matrix.

The ASTM Specification also provides standards for measuring flake size. This is done by comparing a polished microspecimen of the iron at a standard magnification of 100× with a series of standard diagrams (Fig. 3.4b).

Applications of grey iron castings

Grades of grey iron suitable for various types of casting

Unalloyed grey iron is used for a wide variety of castings. The following table indicates typical grades of iron used for making certain types of casting. Table 3.3 is only intended as a guide, casting buyers may specify different grades from those in the table. Some buyers may have their own specifications which do not coincide exactly with international or national specifications.

The production of grey irons

In order to produce castings having the required properties, the following factors must be controlled:

the charge to be melted;
the method of melting;
treatment of the liquid metal before casting.

When deciding on a chemical composition for any casting, the effects of the main constituents have to be considered. These can be summarised briefly as follows:

Carbon. The effect of carbon must be considered together with that of silicon and phosphorus and the concept of carbon equivalent value has already been discussed. CEV has a major effect on the strength and hardness of the casting (Figs 3.2, 3.3). CEV also affects the casting properties of the iron. Irons having CEV below about 3.6% become difficult to cast sound, because liquid shrinkage occurs and feeding is necessary to achieve complete soundness.

Silicon. Next in importance to carbon, with regard to the properties of iron, is silicon. Silicon is a graphitising element, high silicon irons, over about 1.6%Si, tend to be graphitic, while low silicon irons are mottled or white. Silicon contents of grey iron are generally around 2.0%.

Manganese. Manganese is necessary to neutralise the effect of sulphur in

Table 3.3 Typical grades of grey iron and their use

Class of casting	Grade of grey iron used (tensile strength N/mm²)
Air-cooled cylinders	200–250
Agricultural machinery	200–250
Bearing caps	200
Bearing housings	200
Boilers (central heating)	150–200
Brake cylinders	200
Brake discs	200
Brake drums	200
Clutch housings	200
Clutch plates	200
Cooking utensils	150–200
Counterweights	100–150
Cylinder blocks	200–250
Cylinder heads	200–250
Diesel	250
Electric motor stator cases	200
Drain covers, gulleys, etc.	150
Flywheels	200
Gear boxes	200–250
Ingot moulds	100–150
Lathe beds	100
Machine tool bases	200
Manifolds	200–250
Paper mill drier rolls	150–200
Piano frames	150–200
Rainwater pipes, gutters	150
Refrigerator compressors	250
Tractor axles	150–200
gear boxes	200
Valves, low pressure (gas)	150–200
water	200
hydraulic	250
Water pumps	200

iron. Without sufficient Mn, iron sulphide forms during solidification and deposits around grain boundaries where it renders the metal hot-short and likely to produce cracked castings. If sufficient Mn is present, MnS forms as the molten metal cools and floats out of the metal into the slag layer.

The formula: $Mn\% = 1.7 \times S\% + 0.3\%$ represents the amount of Mn needed to neutralise the sulphur, although the sulphur should not be allowed to exceed about 0.12%. In addition to combining with sulphur, Mn is a pearlite stabiliser and it increases the hardness of the iron. However, it is not primarily used for strengthening because it can affect nucleation adversely. Mn levels in most irons are typically 0.5–0.8%.

Sulphur. Sulphur is generally harmful in grey iron, and should be kept to below 0.12%. Even if the sulphur is adequately neutralised with Mn, if the high S slag is trapped in the casting, it can cause blowhole defects beneath the casting skin.

Phosphorus. P has limited solubility in austenite and so segregates during solidification forming phosphides in the last areas to solidify. While P increases the fluidity of all cast irons, it makes the production of sound castings more difficult and the shock resistance of the iron is reduced. For most engineering castings, phosphorus should be kept below 0.12%, but up to 1.0% may be allowed to improve the running of thin section castings where high strength is not required.

Chromium. Added in small amounts, Cr suppresses the formation of free ferrite and ensures a fully pearlitic structure, so increasing hardness and tensile strength. Too much Cr causes chill at the edges of the casting, reducing machinability seriously. Cr up to about 1% may be added to grey irons used for special purposes, such as camshafts, where chills are often used to create wear resistant white iron on the cam noses.

In addition to the above elements, the effect of certain impurities in the iron must also be considered:

Copper. Cu increases tensile strength and hardness by promoting a pearlitic structure and reducing free ferrite. It reduces the risk of chill in thin sections. Up to about 0.5% Cu may arise from the presence of Cu as a tramp element in steel scrap.

Tin. Sn has a similar effect to that of copper, though smaller amounts are effective. Up to 0.1%Sn ensures a fully pearlitic matrix and reduces free ferrite. Like copper, small amounts of tin may arise from steel scrap.

Lead. The presence of lead in grey iron, in amounts as low as 0.0004%, can cause serious loss of strength through its harmful effect on the structure of flake graphite. Lead contamination in cast iron comes usually through the inclusion of free-cutting steel in the scrap steel melting charges, though it can also arise from certain copper alloys such as gun-metal.

Aluminium. Contamination of automotive steel scrap by light alloy components is the usual source of Al in iron. Levels of 0.1% Al may occur. Al promotes hydrogen pick-up from sand moulds and may cause pinhole defects in castings.

Nitrogen. Nitrogen above about 0.01% causes blowholes and fissures in iron castings, heavy section castings are most seriously affected. Nitrogen usually arises during cupola melting, particularly if high steel charges are used, but the use of moulds and cores containing high nitrogen resins can also cause problems. Addition of 0.02–0.03% titanium neutralises the effect of nitrogen.

Chapter 4
Melting cast irons

Introduction

Iron foundries require metal of controlled composition and temperature, supplied at a rate sufficient to match the varying demands of the moulding line. The metallic charge to be melted consists usually of foundry returns, iron scrap, steel scrap and pig iron with alloying additions such as ferrosilicon. The charge is usually melted in a cupola or in an electric induction furnace. Gas-fired or oil-fired rotary furnaces can also be used, but their use is less common.

Cupola melting

The cupola (Fig. 4.1) is the classical iron melting unit and is still the most widely used primary melting unit for iron production due to its simplicity, reliability and the flexibility in the quality of charge materials that can be used because some refining of undesirable elements such as zinc and lead can be achieved. While the cupola is an efficient primary melting unit, it does not adapt easily to varying demands, nor is it an efficient furnace for superheating iron. For this reason it is often used in conjunction with an electric duplexing furnace.

The simplest form is the cold blast cupola which uses ambient temperature air to burn the coke fuel. The metal temperature that can be achieved is normally from 1350 to 1450°C but higher temperatures can be achieved through the use of divided blast (as in Fig. 4.1) or oxygen enrichment. The refractory linings of cold blast cupolas have a short life of less than 24 hours, so cupolas are operated in pairs, each used alternately while the other is re-lined.

In hot blast cupolas (Fig. 4.2), the exhaust gases are used to preheat the blast to 400–600°C, reducing coke consumption and increasing the iron temperature to more than 1500°C. They may be liningless or use long life refractories giving an operating campaign life of several weeks.

'Cokeless' cupolas (Fig. 4.3), have been developed in which the fuel is gas or oil with the charge supported on a bed of semi-permanent refractory spheres. They have advantages of reduced fume emission.

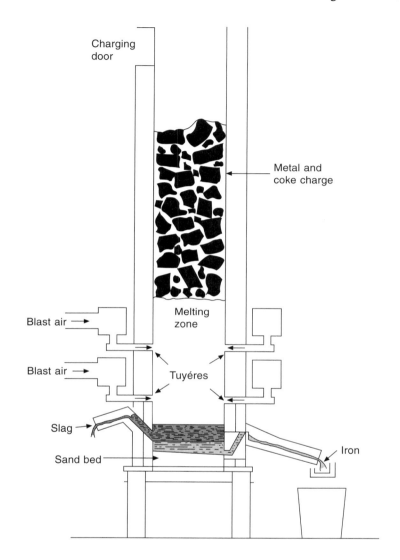

Figure 4.1 *Section through a cupola (From ETSU Good Practice Case Study 161; courtesy of the Department of the Environment, Transport and the Regions.)*

Cold blast cupola operation

The cupola is charged with:

1. coke, the fuel to melt the iron;
2. limestone, to flux the ash in the coke etc.;
3. metallics, foundry scrap, pig iron, steel and ferroalloys;
4. other additions to improve the operation

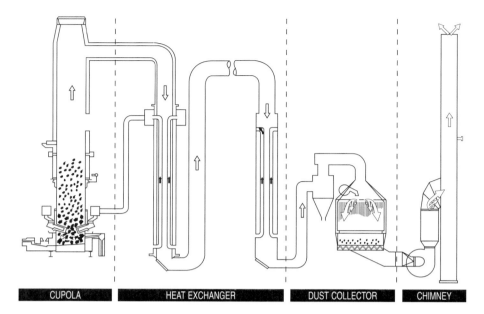

Figure 4.2　*Hot blast cupola. (From ETSU Good Practice Case Study 366; courtesy of the Department of the Environment, Transport and the Regions.)*

The cupola is blown with air to combust the coke and the air flow controls the melting rate and metal temperature. The output of a cupola depends primarily on the diameter of the shaft of the furnace and on the metal/coke ratio used in the charge. Table 4.1 summarises the operating data for typical cold blast cupolas.

A useful measure of the efficiency of operation of a cupola is the 'Specific Coke Consumption' (SSC) which is

$$\frac{\text{Annual tonnage of coke} \times 1000}{\text{Annual tonnage of metallics charged}} = \text{SSC (kg/tonne)}$$

This takes into account both charge coke and bed coke. When the cupola is operated for long enough campaigns, the amount of coke used to form the bed initially can be ignored. However, as the melting period decreases, the role of the cupola bed becomes more important. Table 4.2 summarises data from 36 cupola installations in the UK in 1989. This table provides a useful reference against which the operation of any cold blast cupola can be compared.

Coke

The performance of the cupola is highly dependent on the quality of the coke used. Typical foundry coke has the following properties:

Moisture	5% max.
Ash	10% max.

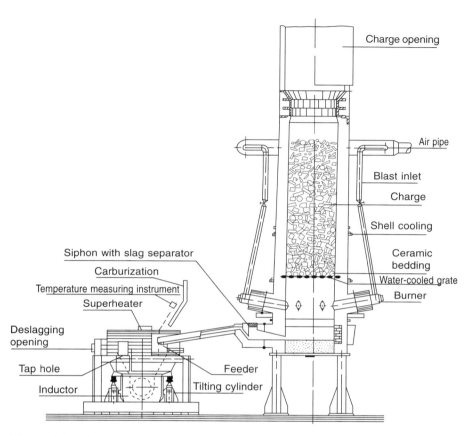

Figure 4.3 *Schematic diagram of a cokeless cupola in a duplex system.* (*From R.F. Taft,* The Foundryman, 86, *July 1993 p. 241.*)

Volatiles	1% max.
Sulphur	1% max.
Mean size	100 mm
Undersize	<5% below 50 mm

The coke size directly affects coke consumption per tonne of iron melted and also the melting rate. Optimum cupola performance is achieved with coke in the size range 75–150 mm, if smaller coke is used, metal temperature is reduced and a higher blast pressure is needed to deliver the required amount of air to the cupola. Increasing the size of coke above about 100 mm has no beneficial effect, probably because large pieces of coke tend to be fissured and break easily during charging and inside the cupola.

Coke usage in the cold blast cupola is typically 140 kg per tonne of iron melted (this is an overall figure including bed coke), it is usual to charge coke at the rate of about 10–12% of the metal charged, but the exact amount used depends on many factors such as tapping temperature required, melting rate and the design of the cupola, see Table 4.2.

Table 4.1 Cupola operation data

Metric units

Diameter of melting zone (cm)	Melting rate (tonnes/h) metal : coke ratio		Blast			Typical charge (kg) at 10:1 coke rate			Bed height above tuyeres (cm)	Shaft height (m) tuyeres to charge door sill
	10:1	8:1	rate m³/h	pressure cm H_2O	kPa	coke	iron	limestone		
50	1.97	1.57	1340	104	10.2	20	200	7	100	2.5
60	2.84	2.46	1940	107	10.5	28	284	9	100	3.0
80	5.11	4.36	3450	114	11.2	51	510	17	105	3.0
100	7.99	6.83	5380	119	11.7	80	800	26	105	3.5
120	11.50	9.79	7750	130	12.7	115	1150	38	110	4.0
140	15.60	13.33	10 600	137	13.4	157	1570	52	110	4.0
160	20.44	17.41	13 800	147	14.4	200	2040	67	110	4.5
180	25.88	22.05	17 450	157	15.4	260	2590	85	115	5.0
200	31.95	27.22	21 550	175	17.2	320	3200	106	115	5.0

Table 4.1 (Continued)

Imperial units

Diameter of melting zone (inches)	Melting rate (ton/h) metal:coke ratio		Blast		Typical charge (lbs) at 10:1 coke rate			Bed height above tuyeres (inches)	Shaft height (feet) tuyeres to charge door sill
	10:0	8:1	rate cfm	pressure in. w.g.	coke	iron	limestone		
18	1.6	1.3	665	40	36	360	12	38	16
24	2.9	2.5	1180	42	65	650	21	39	16
30	4.5	3.9	1840	44	100	1000	33	41	16
36	6.6	5.6	2650	46	150	1500	50	41	19
42	9.0	7.6	3620	48	200	2000	66	42	19
48	11.7	10.0	4720	51	260	2600	86	42	22
54	14.8	12.6	5950	53	330	3300	109	43	22
60	18.3	15.6	7360	57	410	4100	135	43	22
66	22.1	18.8	8900	60	500	5000	165	44	22
72	26.5	22.5	10 650	63	594	5940	196	44	22
78	31.1	26.4	12 500	69	697	6970	230	44	22
84	36.1	30.6	14 500	75	809	8090	267	45	22

The above figures represent good average practice and are intended to act as a rough guide only.

Table 4.2 Data for cupolas in the UK (1989)

Melt rate (t/hr)	Cupola dia. (inches)	Water cooled?	Tapping temp. (°C)	Melt period (hrs)	Bed coke (kg)	Restore coke (kg)	SCC (kg/tonne)	Type of metal	% Coke charge	Divided blast
1.5	22.0	no	1400	1.5	100		407	grey	12.00	
2.0	36.0		1500	3.5	500		252	grey	14.00	
3.0	30.0	no	1300	3.0	500		260	grey	4.00	
3.0	32.0		1340	2.0	600	150	216	grey	8.9	
3.0	32.0	no	1450	7.0	500	60	133	malleable	11.00	
3.0	36.0	yes	1350	3.0	400	150	313	grey	16.00	
3.0	22.0			2.0	100		267	grey	20.00	
3.0	30.0	yes	1450	9.0	210		100	grey		
3.0	26.0		1550	2.0			242	grey		
3.0	30.0	no	1475	7.5	490	0	214	malleable	16.00	
3.0	30.0	no	1430	2.0	400		150	grey	7.20	
3.0	30.0	no	1480	7.0	700	140	208	grey	15.00	
3.5	33.0	yes	1300	3.0	380		212	grey	18.00	
4.0	33.0	no	1450	6.0	1750			grey	9.00	
4.0	35.0	no	1450	3.0	600		217	grey	12.00	x
4.0	36.0	no	1470	3.5	650	150	105	grey	7.00	
4.0	30.0	no	1500	5.0	840	500	211	grey	12.00	
4.0	34.0	no	1460	3.0	375		127	grey	12.4	

(Contd)

Table 4.2 (Continued)

Melt rate (t/hr)	Cupola dia. (inches)	Water cooled?	Tapping temp. (°C)	Melt period (hrs)	Bed coke (kg)	Restore coke (kg)	SCC (kg/tonne)	Type of metal	% Coke charge	Divided blast
4.5	31.0		1550	10.0	1000	86	164	grey	13.50	
5.0	32.0	no	1530	8.0	600		129	grey	12.5	
5.0	33.0	no	1550	4.0	750		243	grey	16.0	x
5.0	36.0	no	1470	8.0	1750	250	178	grey	17.80	x
5.0	36.0	no	1500	2.0	420		144	grey	100.00	
5.0	39.0	no	1460	8.0	450	150	171	grey	11.00	
8.0	42.0	no	1500	4.0	1500		153	grey	11.00	x
8.0	53.9									
8.0	48.0	yes	1550	9.0	900	363	140			
8.0	38.0	no	1490	7.0	1100		115	grey	8.64	x
8.0	48.0	no	1440	8.0	1500		138	grey	wide range	
9.0	42.0	yes	1500	33.0	2000	180	121	grey	10.60	x
9.5	48.0	yes	1460	8.0	1000	500	129	grey	10.0	x
10.0	52.0	no	1450	8.0	1800	400	154	grey	13.10	x
10.0	48.0	yes	1500	15.0	1800	300	138	grey	13.70	x
12.0	48.0		1520	8.5	1800	300	112	grey	9.50	x
12.0	43.0	yes	1530	20.0	1200	200	104	malleable	9.50	x
20.0	72.0	yes	1550	336.0	4000		169	duct	16.50	x

From: Coke consumption in iron foundry cupolas, *Energy of Consumption Guide 7, November 1990*, reproduced by permission of the Energy Efficiency Office of the Department of the Environment.

Fluxes

Fluxes are added to the cupola charge to form a fluid slag which may easily be tapped from the cupola. The slag is made up of coke ash, eroded refractory, sand adhering to scrap metal and products of oxidation of the metallic charge. Limestone is normally added to the cupola charge, it calcines to CaO in the cupola and reacts with the other constituents to form a fluid slag. Dolomite, calcium–magnesium carbonate, may also be used instead of limestone.

The limestone (or dolomite) should contain a minimum of 96% of $CaCO_3$ (and $MgCO_3$) and should be in the size range 25–75 mm.

The amount of the addition is dependent on the coke quality, the cleanliness of the charge and the extent of the lining erosion. Normally 3–4% of the metallic charge weight is used. Too low an addition gives rise to a viscous slag which is difficult to tap from the furnace. Too high an addition will cause excessive attack on the refractory lining. When the coke bed is charged, it is necessary to add around four times the usual charge addition of limestone to flux the ash from the bed coke.

Other fluxes may also be added such as fluorspar, sodium carbonate or calcium carbide. Pre-weighed fluxing briquettes, such as BRIX, may also, be used. BRIX comprises a balanced mixture of fluxing agents which activates the slag, reduces its viscosity and produces hotter, cleaner reactions in the cupola. This raises carbon content, reduces sulphur and raises metal temperature.

Correct additions of flux are essential for the consistent operation of the cupola and care should be taken to weigh the additions accurately.

The metallic charge

Table 4.3 gives the approximate metal compositions needed for the most frequently used grades of grey iron. (Data supplied by CDC.)

Table 4.3   Metal composition needed to produce the required grade of grey iron

Grade	150	200	250	300	350
Total carbon (%)	3.1–3.4	3.2–3.4	3.0–3.2	2.9–3.1	3.1 max
Silicon (%)	2.5–2.8	2.0–2.5	1.6–1.9	1.8–2.0	1.4–1.6
Manganese (%)	0.5–0.7	0.6–0.8	0.5–0.7	0.5–0.7	0.6–0.75
Sulphur (%)	0.15	0.15	0.15 max	0.12 max	0.12 max
Phosphorus (%)	0.9–1.2	0.1–0.5	0.3 max	0.01 max	0.10 max
Molybdenum (%)				0.4–0.6	0.3–0.5
Cu or Ni (%)					1.0–1.5

Note: Copper may partially replace nickel as an alloying addition

Metallic charge materials

The usual metallic charge materials are:

Return scrap: runners, risers, scrap castings etc. arising from the foundry operation. Care must be taken to segregate each grade of returns if the foundry makes more than one grade of iron.

Pig iron: being expensive, the minimum amount of pig iron should be used. Use of pig iron is a convenient way of increasing carbon and silicon content. Special grades of pig iron having very low levels of residual elements are available and they are particularly useful for the production of ductile iron.

Steel scrap: is normally the lowest cost charge metal, it is used for lowering the total carbon and silicon contents.

Bought scrap iron: care must be taken to ensure that scrap of the correct quality is used, particularly for the production of the higher strength grades of iron.

Harmful materials

Care must be taken to ensure that contaminants are not introduced into the iron. The most common harmful elements are:

Lead, usually from leaded free-cutting steel scrap.
Chromium, from stainless steel.
Aluminium, from aluminium parts in automotive scrap.

Size of metallic charge materials

Thin section steel scrap (below about 5 mm) oxidises rapidly and increases melting losses. On the other hand, very thick section steel, over 75 mm, may not be completely melted in the cupola. Metal pieces should be no longer than one-third of the diameter of the cupola, to avoid 'scaffolding' of the charges.

Ferroalloys

Silicon, manganese, chromium, phosphorus and molybdenum may all be added in the form of ferroalloys. In some countries, Foseco supplies briquetted products called CUPOLLOY designed to deliver a specific weight of the element they introduce, so that weighing is unnecessary.

Ferrosilicon in lump form, containing either 75–80% or 45–50% Si may be used. Ferromanganese in lump form contains 75–80% Mn. Both must be accurately weighed before adding to the charge.

Pig irons

Typical pig iron compositions are given in Table 4.4. Refined irons for foundry use are normally made in a hot blast cupola from selected scrap, they may contain copper, tin, chromium and other alloy elements. Base irons for ductile (s.g., nodular) iron production are made from specially pure ores, and have very low residual element contents. They are available in a range of specifications.

Table 4.4   Foundry pig iron

Grade	Typical composition				
	TC(%)	Si(%)	Mn(%)	S(%)	P(%)
Blast furnace irons	3.4–4.5	0.5–4.0	0.7–1.0	0.05 max	0.05 max
Refined irons	3.4–3.6	0.75–3.5	0.3–1.2	0.05 max	0.1 max
Ductile base irons	3.8	0.05–3.0	0.01–0.20	0.02 max	0.04 max

Purchased cast iron scrap is available in a number of grades, typical compositions are shown in Table 4.5.

Table 4.5   Cast iron scrap

Type	Typical composition				
	TC(%)	Si(%)	Mn(%)	S(%)	P(%)
Ingot mould scrap	3.5–3.8	1.4–1.8	0.5–1.0	0.08	0.1
Heavy cast iron scrap	3.1–3.5	2.2–2.8	0.5–0.8	0.15	0.5–1.2
Medium cast iron scrap	3.1–3.5	2.2–2.8	0.5–0.8	0.15	0.5–1.2
Automobile scrap	3.0–3.4	1.8–2.5	0.5–0.8	0.15	0.3 max

Typical charges needed to produce the most frequently used grades of iron are given in Table 4.6.

Table 4.6   Typical furnace charges

Grade 150	Grade 200	Grade 250
25% pig iron	30% low P pig iron	25% low P pig iron
40% foundry returns	35% foundry return	35% foundry returns
30% bought cast iron	20% low P cast iron scrap	15% low P scrap
5% steel scrap	15% steel scrap	25% steel scrap

Cupola charge calculation

In a normally operated, acid cold blast cupola, the composition of the metal tapped can be predicted with reasonable accuracy from the composition of the furnace charge. The tendency is for the total carbon to attain the eutectic equivalent. If the quantity charged is above this value, a loss may be expected. On the other hand, where the charge contains less than the eutectic value, the trend is towards a carbon pick-up. The exact amount of carbon change must be established by experience for a particular cupola operation, but the following 'Levi equation' is a good starting guide.

$$\text{TC\% at spout} = 2.4 + \frac{\text{TC\% in the charge}}{2} - \frac{\text{Si\% + P\% at spout}}{4}$$

Silicon is always lost in the cupola, generally a loss of 15% of that charged may be assumed, but higher losses may occur if high steel charges are used. Manganese losses are usually about 25%. Phosphorus changes little. Sulphur always increases due to pick-up from the coke, but the precise amount cannot be predicted and must be based on experience.

Based on these guide lines, a calculation may be made as follows:

To make a Grade 250 iron with the composition:

TC	Si	Mn	P
3.2	1.7	0.7	0.1

Material	Amount charged (%)	Composition				Contribution to charge (%)			
		TC	Si	Mn	P	TC	Si	Mn	P
Low-P pig iron	25	3.0	3.0	1.0	0.1	×0.25	0.75	0.75	0.25	0.03
Grade 250 returns	35	3.2	1.7	0.7	0.1	×0.35	1.12	0.60	0.25	0.04
Low-P scrap iron	15	3.2	2.2	0.8	0.15	×0.15	0.48	0.33	0.12	0.02
Steel scrap	25	0.1	0.1	0.3	0.03	×0.25	0.03	0.03	0.08	–
Ferromanganese	0.3			75		×0.003			0.23	
Total							2.38	1.71	0.93	0.09
Changes during melting		Si loss 15%						−0.26		
		Mn loss 25%							−0.23	
Addition at spout		70% ferrosilicon						+0.25		

Expected composition T.C. = 2.4 + 2.38/2 − (1.45 + 0.09)/4
= 3.2
Si = 1.70
Mn = 0.70
P = 0.09

Calculations such as the above example, should only be used as a guide. The precise carbon pick-up and silicon losses achieved depend on factors such as coke quality, metal temperature, melting rate etc. Experience will enable more accurate predictions to be made. A number of computer programs are available which carry out the calculations rapidly and enable 'least cost' charges to be selected.

Cupola output

The maximum output from a cupola is determined primarily by the shaft diameter, Table 4.1. In normal use, it is necessary to be able to vary the output to match the requirements of the moulding line. This is done by varying the blast rate. Increasing or reducing the air supplied to the cupola burns more or less coke and increases or reduces the melting rate. Unfortunately, changing the blast rate also changes the temperature of the metal, and to some extent, its composition. This is one of the main drawbacks of the cupola as a melting furnace. Another problem is that the only way of changing the composition of the liquid iron is to change the make-up of the charge, and it usually takes around an hour before the change is seen at the tap-hole. To overcome these difficuties, it is common practice to tap the cupola into an electric holding furnace where the temperature and composition can be accurately controlled, and variations in metal demand can be accommodated.

Emissions from cupolas

Exhaust gases from cupolas are hot and contain dust, grit and SO_2 gas. For many years, the emissions permitted from cold blast cupolas were readily achieved by the use of simple wet arresters. The cupola gases pass through a curtain of water which removes the grit particles, absorbs up to half of the SO_2 but does not remove dust. Present day environmental regulations in most countries impose increasingly strict limitations on the dust emissions permitted from cupolas, requiring additional dust-arresting plant to be fitted. Wet scrubbers, bag filters and electrostatic precipitators can be used.

There are two types of wet scrubbers: venturi scrubbers and disintegrators. Venturi scrubbers rely on the pressure drop across a restricted throat and disintegrators on the wetting and agglomeration of dust particles by the action of water carried by a rapidly spinning rotor. Capital cost and running costs, power and maintenance, are high.

Dry bag filters are capable of achieving lower emission levels than wet scrubbers. The gases must be cooled before filtration making capital costs higher than wet scrubbers but running costs may be lower.

Electrostatic precipitators are efficient but are expensive and require specialised maintenance, they are uncommon on foundry cupolas.

The long campaign hot blast cupola

Hot blast cupolas were, until recently, only considered economical for foundries with large continuous requirements for molten iron. Hot blast cupolas are operated on long campaigns, many with unlined, water-cooled steel shells. Independently fired blast systems have been used but they have high fuel costs and have now been largely abandoned. There is now renewed interest in the long campaign hot blast cupola, with recuperative systems using the heat from combusting the cupola offtake gases to heat the blast (Fig. 4.2). In part, the change has come about because of environmental concerns. Cold blast cupolas, in the UK and elsewhere, have in the past been allowed to operate with simple, low cost emission control. Environmental controls are now becoming more stringent, requiring high efficiency filtration of cupola offtake gases. This generally demands combustion then cooling of the gases prior to filtration. Rather than waste this heat, more foundries are turning to the hot blast cupola.

The upper section of the cupola is lined with refractory, while the melting zone may be liningless (having a water-cooled shell) or it may use a high quality backing lining with a replaceable inner lining. The liningless cupola can be operated for several weeks without dropping the bottom. The refractory lined cupola is usually operated for a week without replacing refractories. Combustion of the offtake gases is maintained by introducing air into the shaft below the charging door together with a gas-fired afterburner which automatically ignites if the temperature of the gases falls too low. The offtake gases are drawn from the cupola through a recuperator which preheats the incoming blast air to around 500°C. The blast air is enriched with 1.5–2.0% oxygen. The waste gases are cooled to 175°C before passing through a dry bag filter prior to discharge to atmosphere.

Tapping temperatures of 1530°C are achieved. The main savings over conventional cold blast cupola practice is found in the reduced coke consumption. Savings of up to 30% of coke usage are claimed. Long campaign cupolas can be designed for economical operation from 10 tonnes/hr upwards. The long campaign hot blast cupola is considered by many to be the most economical method of melting grey iron for foundries.

The cokeless cupola

This is a continuously melting tower furnace in which the metallic charge is supported on a water-cooled grate on which is a bed of carbonaceous refractory spheres. Heat for melting is provided by gas (or oil) burners (Fig. 4.3). Superheating of the liquid iron is performed by the heated refractory spheres and carbon can be added by injecting a suitable recarburiser into the well of the cupola. Eliminating the coke eliminates sulphur pick-up, making the cokeless cupola suitable for the production of base iron for ductile iron production. It also eliminates the main source of atmospheric pollution. The cokeless cupola retains the advantages of cupola melting: continuous operation, ability to accept a wide range of raw materials including wet, oily and contaminated scrap and some refining which removes harmful elements such as lead and zinc.

The cokeless cupola is particularly attractive in countries where good quality foundry coke is not available. The most efficient way of using the cokeless cupola is to tap at around 1350–1400°C into an electric duplexing furnace where temperature and composition are controlled. This practice reduces gas consumption to about 55 m^3/tonne of metal melted and greatly reduces the consumption of the refractory spheres of the bed to around 1 kg/tonne of metal melted. Cokeless cupolas with capacity from 5–15 tonnes/h are in use.

Electric melting

Electric melting in the form of arc, induction and resistance furnaces is used increasingly, both for primary melting and holding of liquid iron. Induction furnaces are the most popular, there are two basic types, the channel furnace and the coreless induction furnace. Compared with cupola melting, induction furnaces offer the following advantages:

The ability to produce iron of closely controlled composition.
The ability to control tapping temperature precisely.
A wide range of charge materials can be used.
Environmental pollution is much reduced.
The stirring effect of induction power rapidly incorporates additions.

The channel furnace

The channel furnace consists of an upper vessel, holding the bulk of the charge material, with an inductor bolted on the underside (Fig. 4.4).

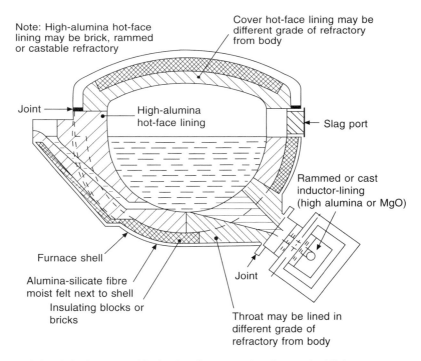

Figure 4.4   *A bath channel induction furnace, showing typical lining arrangement. (Reproduced by courtesy of CDC.)*

All the power is induced within the inductor, heating a loop of molten metal which transfers its heat to the main body of the charge by convection and induction forces. The channel furnace will only work providing this loop is maintained 24 hours per day. The temperature of the metal in the loop is higher than that of the metal in the main vessel, which limits the operating temperature of a channel furnace, since a high loop temperature shortens the life of the refractory of the loop. This means that their use for melting is restricted to low temperature metals, such as brass or aluminium. Channel furnaces are frequently used to duplex iron from a cupola, that is, the cupola is used as the primary melting unit and continuously delivers liquid iron to a large (say, 20 tonnes) channel furnace where compositional variations are corrected and the temperature is maintained at the required value. This means that the varying demand from the foundry can be met with metal of precisely the required specification, which cannot be achieved with a cupola alone.

The disadvantage of the channel furnace is the necessity to operate it for 24 hours per day, and the problems which arise if the refractories of the induction loop fail.

The coreless induction furnace

In a coreless furnace, the coil surrounds the entire charge (Fig. 4.5). The mass of refractory is much less than in the channel furnace while the shape is a simple hollow cylinder. Hence coreless furnaces are much simpler and

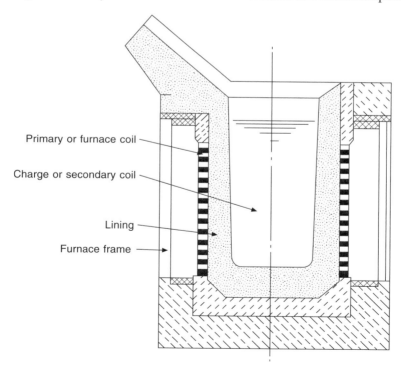

Primary or furnace coil

Charge or secondary coil

Lining

Furnace frame

Figure 4.5   *Section through a coreless induction furnace.* (*From Jackson, W.W. et al. (1979)* Steelmaking for Steelfounders, *SCRATA; reproduced by courtesy of CDC.*)

less costly to reline, although they require more frequent relines than the channel furnace.

The coreless furnace can be designed to operate at any frequency from 50 Hz upwards. Induction heating of liquid metal causes a stirring effect in the metal. The lower the frequency of the primary current, the more intense is the stirring. Therefore in a mains frequency furnace operating at 50 or 60 Hz, the turbulence is greater than in one operating at a higher frequency. Because of the high turbulence, the power input to a mains frequency furnace is restricted to around 250 kW per tonne of capacity. With higher frequencies, the power density can be increased to three or four times this level.

The frequency of operation also affects current penetration. The current induced in the metal charge is a maximum at the surface, reducing at greater depths. The higher the frequency, the less the effective penetration depth. This means that at 50 Hz, the smallest ferrous charge piece that can be efficiently heated is around 450 mm diameter. The smallest practicable furnace that can hold such a piece is 750 kg in capacity. Mains frequency furnaces are not effective in sizes smaller than this. On the other hand, at 10 kHz charge pieces less than 10 mm diameter can be heated, so furnaces as small as 5 kg capacity can be used.

For most foundry applications, furnaces operating in the range 250–3000 Hz are used, such furnaces:

allow high power densities giving high melting rates without excessive stirring;
can be emptied completely after each melt and restarted with virtually any size of scrap;
have short start-up time and offer flexibility for alloy changes;
allow high production rates from small furnaces;

The development of the solid state inverter in the early 1970s allowed the possibility of a cheap and reliable form of frequency converter. Silicon controlled rectifiers (SRCs) convert AC to DC and back to AC at the desired frequency. Once these had been fully developed, the medium frequency induction furnace replaced the earlier mains frequency and triple frequency furnaces. So whereas in the past, choice of frequency was invariably a compromise because of the limited range possible, now the correct unit for the application can be supplied. One supplier (Inductotherm) for example, has supplied:

70 Hz	for brass swarf
100 Hz	for aluminium scalpings
100–150 Hz	for cast iron borings
250 Hz	for Al extrusion scrap
250 Hz	for iron from foundry returns and steel scrap
500 and 1000 Hz	for steel melting and for melting wet cast iron borings
1000 and 3000 Hz	for melting a wide variety of copper alloys
3000 Hz	for investment casting
10000 Hz	for the jewellery trade

For iron melting furnaces having a melt capacity of around 2–20 tonnes, frequencies of 200–1000 Hz are usually used, being powered at up to 750 kW per tonne of capacity. Such furnaces normally have a charge-to-tap time of about 1–2 hours and produce metal at the rate of 2–10 tonnes per hour.

Energy consumption

A figure of about 500 kWh/tonne is attainable when using a high-powered medium frequency furnace in a melting role, i.e. when holding periods are minimised. The energy used can be broken down approximately for a furnace having the following characteristics (from C.F. Wilford, *The Foundryman*, 1981 p. 153):

Furnace capacity	4 tonnes
Installed power	3000 kW
Frequency	500 Hz
Standing heat losses (lid on) averaged over complete melt cycle	47 kW
Power supply efficiency (to furnace coil)	96%
Coil efficiency	80%
Bath diameter	930 mm

Calculated energy consumption

Theoretical energy to melt 4 tonnes of charge to 1500°C		1480 kWh
Energy consumed (allowing for efficiency factors)		1927
Furnace heat losses over one cycle (lid on)		40
Charging time (lid off)	30 min	
Extra energy loss incurred with lid off		10
Time for deslagging, temperature measurement and analysis (lid off)	5 min	
Extra energy incurred with lid off		16
Energy consumed to replace lid off losses		34
Total energy consumed (for 4 tonnes)		2001
	kWh/t	500

The above figures are theoretical, surveys of actual foundry installations show that figures of 520–800 kWh/tonne are common, the variation being due to individual melting practice such as the rate at which the pouring line will accept molten metal and whether furnace lids are used effectively. Attention to energy saving measures should allow figures of 550–650 kWh to be achieved.

Charge materials

The maximum dimension of a piece of charge is around one-third of the crucible diameter. If larger, there is a danger of bridging. The value may be exceeded when a long piece of charge is fed in a controlled manner into the crucible. Charge bridging can also be a problem when melting charges of cast iron borings. It is difficult to generate circulating electric currents in a cold charge having high electrical resistance between individual pieces (as found with oxidised borings), so a large sintered mass may form which does not easily sink down as melting occurs.

An advantage of the medium frequency furnace is that wet charge components can safely be charged into an empty crucible, eliminating the necessity of using a charge pre-drying stage. However, care must be taken not to charge wet material into a fully molten bath.

Alloy recovery

Addition	Recovery (%)
Carbon (graphite, petroleum coke)	80–88
Silicon (ferrosilicon 75%)	90
Manganese (ferromanganese 75%)	90–95
Chromium (ferrochromium 60–75%)	90
Molybdenum (ferromolybdenum 60–70%)	90
Nickel	95
Copper	95
Sulphur (iron sulphide)	100

Slag removal

Slag is formed in electric melting furnaces from the products of oxidation of the elements in the charge, particularly the iron, silicon and manganese; from refractory erosion and from dirt, sand or rust on the charge. Slag floats on the surface of the metal and must be removed before tapping. Since slag removal is an unpleasant task for the furnace operator, it is advisable to avoid the use of dirty or rusty charge materials. Many slags are liquid and of low viscosity and difficult to collect, they can be coagulated by adding SLAX. At molten metal temperatures the SLAX granules expand and form a low density, high volume crust which mops up the slag, which can then be lifted off the metal with ease, leaving the surface clean. SLAX is based on siliceous minerals and reacts with the slag to increase the silica content, and hence the viscosity. Between 0.07 and 0.2% by weight (0.7–2.0 kg/tonne of metal) of SLAX 10 or SLAX 30 is scattered over the slag on top of the furnace. This is rabbled to form a dry, expanded crust and skimmed off.

Refractories for coreless induction furnaces

Coreless furnaces in iron foundries are usually lined with a silica refractory bonded with boric acid or, preferably boric oxide. The quality of the silica is important, it is mined as quartzite and should have the following approximate composition:

SiO_2	Al_2O_3	Fe_2O_3	CaO	MgO	Alkali
98.9%	0.6%	0.2%	0.1%	0.04%	0.2% max

Correct particle sizing is essential so that the lining can be compacted to as high a density as possible. Typical gradings are:

20%	> 1 mm
20%	0.5–1.0 mm
30%	0.1–0.5 mm
30%	< 0.1 mm

Boric oxide is usually used as the bonding agent, being mixed by the refractory supplier. Around 0.7–0.8% B_2O_3 is used. During the fritting cycle, as the temperature at the hot face increases, the boric oxide dissolves the silica fines, producing a borosilicate glass which fills the interstices between the silica grains and cements them together.

The usual practice with medium frequency furnaces is to coat the copper coil of the furnace with a layer of 'mudding' about 6 mm thick, of a medium to high alumina cement. This remains in place when the hot face is knocked out. Between this and the hot face is a layer of ceramic fibre insulation. The working face is formed by compacting the silica refractory behind a steel former concentrically placed within the coil. Formers are normally constructed from mild steel sheet according to the furnace manufacturer's design. Refractory is poured between the former and the coil and compacted using vibratory ramming tools or manual compaction. Inhalation of silica dust presents a hazard and respirators should be used during installation or wrecking of the lining.

The lining is fritted by slow inductive heating of a metallic charge placed inside the steel former. The heating rate depends on the size of the furnace and the manufacturer's recommendations should be followed. In general, a low heating rate of 50–100°C/h should be used until the temperature reaches 700°C, after which the rate can be increased to 100–200°C/h (the faster rate being possible with furnaces of size 10 tonnes or less). The temperature should be raised to 30–50°C above the normal operating temperature of the furnace, and held for about one hour to complete the fritting operation.

The lining life is very dependent on the particular practice used in the foundry and the type of iron being melted. For example, the high carbon and low silicon contents of most ductile base irons, together with the higher temperatures involved, tends to result in lower lining lives than furnaces used for grey or malleable irons. A 5-tonne furnace melting grey iron should be capable of melting 600–800 tonnes of iron from one lining, but less than half this amount if a ductile base iron is being melted.

If a range of irons is melted, requiring higher temperature than normal, or producing more aggressive slags, it may be necessary to use an alumina lining. These can be used up to 1750°C and have greater corrosion resistance, but the cost is much higher. The installed lining cost is around 2.5 times greater than a silica lining, although the life may be three times longer.

Operating systems

Most iron foundries use two furnace bodies, identical in size, fed from a single power supply with some means of switching the power supply between the two furnaces. This allows a continuous supply of molten metal with one furnace dispensing molten metal while the other is melting the next batch. Switching techniques have been developed to enable a single power supply to provide melting power to one furnace while simultaneously providing holding power to the second so that temperature control can be maintained.

Mechanised charging systems, vibratory conveyors or drop bottom charging buckets are frequently used to ensure maximum furnace utilisation. De-slagging is the most arduous and time consuming operation, back tilting the furnace aids the process.

While most of the ferrosilicon and carburiser are added during furnace charging, some carbon and silicon losses will occur at high molten metal temperatures. Trimming additions of 0.2 to 0.3%C and 0.2%Si are typical during the final stages of the melting process, the stirring action of the medium frequency power allowing rapid solution and consistent metal composition.

Fume extraction

Electric melting plant produces less fume than cupola melting, but the ever increasing stringency of environmental regulations requires fume extraction plant to be fitted. The charging and pouring operations generate the majority of the dust and fume emissions within the melt cycle. Close capture fume hoods or high velocity lip extraction units are specified for fume capture and dry bag filter systems are sufficient to handle the levels of fume that arise.

Shop floor control of metal composition

The carbon and silicon content of unalloyed cast irons can be quickly determined on the shop floor by thermal analysis. A sample of molten iron is poured into a small, expendable test mould about 25 mm diameter and 65 mm deep made from resin bonded sand and coated with tellurium to ensure that the sample freezes white. The test mould also contains a thermocouple connected to a temperature recorder. As the sample solidifies, the temperature recorder plots a cooling curve which displays the 'liquidus arrest' when the sample first starts to freeze, then the 'eutectic arrest' when freezing is complete. The liquidus arrest measures the carbon equivalent liquidus value (CEL) given by:

$$CEL = \%C + \frac{\%Si}{4} + \frac{\%P}{2}$$

Note that CEL is not the same as CEV, $\%C + (\%Si + \%P)/3$.

The eutectic arrest temperature, on an unalloyed iron of low P%, is a measure of the silicon content of the iron. Thus from the two arrest temperatures, the carbon and the silicon content can be calculated. Simple calculators are available to enable the C% and Si% to be read, or digital meters are available which display the C% and Si% directly. The carbon content can be determined with an accuracy of ±0.05%. The silicon measurement has an accuracy of about ±0.15%

To achieve good results:

the sample must be poured at a high enough temperature to give a well-defined liquidus arrest. This means that the sampling spoon must be preheated;

the iron should not be inoculated before testing;

the sample must solidify white.

Special expendable sample moulds are available for the purpose.

Chapter 5

Inoculation of grey cast iron

Introduction

In order to achieve the desired mechanical properties in iron castings, the liquid iron must have the correct composition and it must also contain suitable nuclei to induce the correct graphite structure to form on solidification. The liquid iron must have a suitable 'graphitisation potential', this is determined mainly by its carbon equivalent value, and in particular by the silicon content. It is normal practice to adjust the graphitisation potential by controlling the silicon content. However, the effect of other elements must also be considered. Table 5.1 shows the effect of common alloying elements relative to silicon for concentrations normally found in practice.

Table 5.1 The graphitising and carbide stabilising effect of elements relative to Si

Graphitisers	Carbide stabilisers
C +3.0	Mn −0.25
Ni +0.3	Mo −0.35
P +1.0	Cr −1.20
Cu +0.3	V −1.0 to 3.0
Al +0.5	

From *Cast Iron Technology*, Elliott, R. (1988), Butterworth-Heinemann, reproduced by permission of the publishers.

Example: the effect of 1%Al is approximately equivalent to the graphitising power of 0.5%Si. 1%Cr will neutralise the effect of about 1.2%Si.

Even if iron of the correct chemical analysis is made in the melting furnace, castings having the desired graphite structure will not be produced without the addition of inoculants. Inoculants are alloys added in small amounts to induce eutectic graphite nucleation. Without the presence of suitable nuclei, liquid iron will 'undercool' below the eutectic temperature (Fig. 5.1). Uninoculated grey iron castings will contain:

undercooled forms of graphite, associated with this will be ferrite; cementite in thin sections or close to edges and corners.

Such iron is unlikely to meet tensile and hardness specifications and will be difficult to machine.

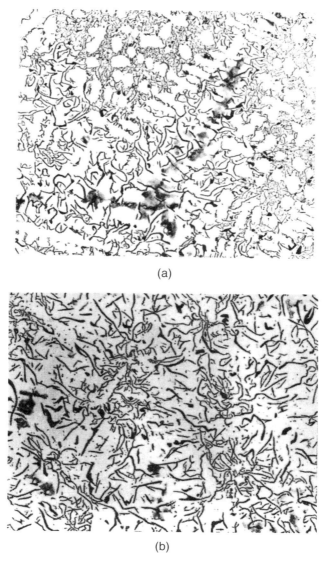

(a)

(b)

Figure 5.1 *Graphite structure of (a) uninoculated grey cast iron (×100) and (b) inoculated grey cast iron (×100). (From BCIRA Broadsheet 161-4; reproduced by courtesy of CDC.)*

There are two main methods of inoculation, ladle and late inoculation. In the former, the inoculant is added either as the liquid iron enters the ladle or just afterwards. Late inoculation refers to treatment after the metal has left the ladle, for example, as it enters the mould (stream inoculation) or by using an insert in the mould (in-mould inoculation). Inoculants reach maximum effectiveness immediately after treatment and fade quickly over a period of 10–20 minutes. It is therefore desirable to inoculate as late as possible before casting.

Inoculants are mostly based on graphite, ferrosilicon or calcium silicide, with ferrosilicon being the most commonly used. Pure ferrosilicon is not effective as an inoculant, it is the presence of minor elements that determine the effectiveness of the product. Graphite itself is a powerful inoculant but it is not effective on low sulphur irons.

Table 5.2   INOCULIN products for inoculation of grey, ductile and compacted graphite irons

INOCULIN product	Active constituents	Use
10	Graphite, alloys containing Si, Ca, Al, Zr	General grey iron, especially where S exceeds 0.08%. Recommended for max. chill reduction and best graphite structures in cupola iron. Does not affect composition of the iron.
25	Ferroalloy containing 65% Si, Ca, Al, Zr and Mn	All types of grey, ductile and CG irons. High solubility even in low temperature metal.
80	Ferroalloy containing 75% Si, Ca, Zr, Al	Powerful inoculant for grey, ductile and CG irons. Good fade resistance.
90	Specially graded INOCULIN 25	Fine powder inoculant for use in MSI 90 Stream Inoculator.
98	Specially graded INOCULIN 80	Graded for late stream inoculation in MSI 90 Stream Inoculator.

In addition to the above range of inoculants, Foseco supplies certain special grades in some countries for particular applications such as low sulphur irons and for ductile pipe manufacture.

Ladle inoculation

The selected grade of INOCULIN for ladle inoculation should always be added to the metal stream when tapping from furnace to ladle, or ladle to ladle. Additions should begin when the ladle is one-quarter full and be completed when the ladle is three-quarters full, so that the last metal merely mixes.

Never put INOCULIN into the bottom of the ladle and tap onto it.

The amount of inoculant needed is governed by several factors. The following rules guide the use of inoculation:

Low carbon equivalent irons require greater amounts of inoculant.
Grey cast irons with less than 0.06% sulphur are difficult to inoculate, specially formulated products may be required.
For a given iron, the thinner the section of casting, the greater the inoculation required.

Electric melted irons require more inoculation than cupola melted irons.
Electric melting will also produce low sulphur contents.
High steel scrap charges will require more inoculation.
Where inoculated iron is held for more than a few minutes after inoculation,
there is a need of a higher level of treatment.

It is therefore difficult to give an accurate estimate of the amount of INOCULIN
which is required for every situation. In general, INOCULIN additions of
0.1–0.5% by weight of metal will be satisfactory for grey cast irons, higher
additions are needed for ductile (SG) irons (see p. 79). Care must be taken
not to over-inoculate grey irons, otherwise problems will arise with shrinkage
porosity due to too high a nucleation level. Many grades of INOCULIN
contain high Si content, so that by adding 0.5% of inoculant, the silicon
content of the iron will be raised by as much as 0.3%, this must be allowed
for by adjusting the Si analysis of the furnace metal.

Control methods

The wedge chill test is a simple and rapid method of assessing the degree of
chill reduction obtained by the use of INOCULIN in grey cast irons. Carried
out on the foundry floor, the wedge test is frequently used as a routine
check even when full laboratory facilities are available. The most common
dimensions for the wedge are illustrated in Fig. 5.2.

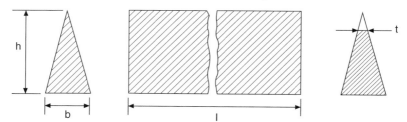

Base (b)		Height (h)		Length (l)	
mm	in	mm	in	mm	in
6	$1/4$	11	$7/16$	57	$2 1/4$
13	$1/2$	22	$7/8$	100	4
19	$3/4$	38	$1 1/2$	127	5
25	1	57	$2 1/4$	127	5

Figure 5.2 *The wedge chill test.*

The wedge is made in a mould prepared from silicate or resin bonded
sand. After pouring, it must be allowed to cool in the mould to a dull red
heat (c. 600°C), after which it can be quenched in water and fractured. The
width at the point where clear chill ceases, *t*, is measured and this gives a
good indication of the need for inoculation and of the effectiveness of an

inoculation process. In general, casting sections should be not less than three times the wedge reading if chill at the edges and in thin sections is to be avoided.

After ladle inoculation, the metal must be cast quickly to avoid inoculant fade.

For certain applications such as continuous casting of iron bar or automatic pouring of castings, inoculant can be added in the form of filled steel wire containing INOCULIN 25 which can be fed into a ladle or the pouring basin of an automatic pouring machine at a computer-controlled rate using the IMPREX Station (see pp. 73, 78). IMPREX wire is available in a range of diameters from 6 mm upwards.

Late stream inoculation

With the increasing number of foundries where castings are made on highly mechanised moulding and pouring lines, the requirements for inoculation are becoming more difficult to meet. Particular difficulties arise with the use of automatic pouring furnaces where conventional ladle inoculation is not possible. A method of carrying out inoculation at the casting stage is needed and this must be consistent and automatic in operation. The MSI 90 Metal Stream Inoculator is intended for use in these conditions.

It is designed to add controlled amounts of inoculant to the liquid cast iron just before it enters the mould. The use of late stream inoculation techniques leads to the virtual elimination of fading. This permits a substantial reduction in the amount of inoculant used. The inoculant addition thereby produces a smaller change in iron composition leading to improved metallurgical consistency. The cost of inoculation is also lower.

The MSI 90 Stream Inoculator consists of two units, Fig. 5.3, a control unit and a dispensing unit linked together by a special cable and air line assembly. The inoculant dispensing cabinet is located in a fixed position over the mould being poured. A storage hopper for the inoculant is mounted above the dispensing cabinet. In the latest version, MSI SYSTEM 90-68E, Fig. 5.4, the flow of inoculant can be regulated either by optical detection of the start and end of iron flow via an optical module and fibre optic system or by connecting the system to the pouring furnace electrical signal used to regulate the flow of liquid iron. The monitoring system checks INOCULIN 90 level, dispensing tube status, inoculant flow, gate status, compressed air and dispensing unit temperature. The monitor can automatically interrupt pouring in the event of malfunction. The control unit is fitted with a printer port allowing records to be kept. The control cabinet is positioned in a secure, easily accessible place and may be some distance from the point of inoculation.

The MSI 90 Stream Inoculator can be operated in conjunction with a variety of types of pouring equipment:

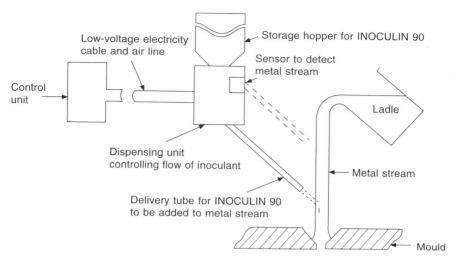

Figure 5.3 *The principles involved in the MSI System 90.*

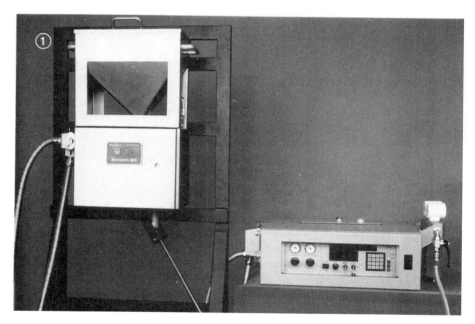

Figure 5.4 *MSI System 90 Type 68E.*

pouring furnaces
ladle transporters
automatic ladle pouring devices
conventional ladles with fixed or variable pouring positions (provided the latter is within a limited radius).

The inoculant used in late stream inoculators must have a number of important features:

It must be a powerful inoculant.
It must be finely divided to ensure free-flowing properties and rapid solution.
It must be very accurately graded, without superfine material which would blow away, or large particles which jam the gate mechanism.
It must dissolve rapidly and cleanly to avoid the presence of undissolved inoculant particles in the castings.

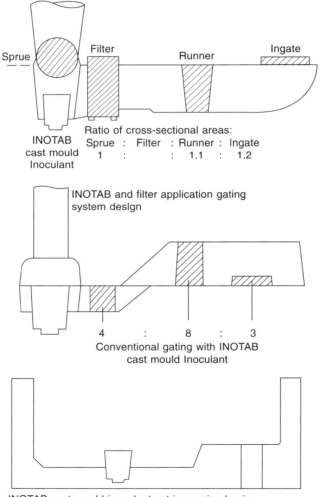

Figure 5.5   *Application of INOTAB cast mould inoculant.*

These requirements are met by INOCULIN 90, specially developed for this purpose. INOCULIN 90 is an inoculating grade of ferroalloy containing balanced proportions of Si, Mn, Al, Ca and Zr, and is an excellent inoculant for grey and ductile irons. INOCULIN 90 should not be used for normal ladle inoculation because of its very fine size grading.

Stream inoculation is very efficient since fading is eliminated. The normal addition rate for grey iron is from 0.03–0.20%, typically 0.1%, much less than would be used for ladle inoculation. For ductile iron, addition rates range from 0.06–0.3%, typically 0.2%.

Mould inoculation

There are several ways in which mould inoculation can be performed: powdered inoculant can be placed in the pouring bush; or it can be placed at the bottom of the sprue. A more reliable method is to use sachets or precast slugs of inoculant in the pouring bush or in the running system (Fig. 5.5).

INOPAK sachets are sealed paper packets containing 5, 10 or 20 g of graded, fast-dissolving inoculant which can be placed in the runner bush, at the top of the sprue or in some other situation where there is a reasonable degree of movement in the metal stream. For most purposes, the addition rate should be 0.1%, i.e. 5 g of INOPAK for each 5 kg of iron poured.

INOTAB cast mould inoculant tablets are designed to be placed in the runner where they gradually dissolve in the metal stream as the casting is poured, giving uniform dissolution. This ensures that inoculation takes place just before solidification of the iron. Application is simple using core prints to locate the INOTAB tablet.

INOTAB tablets are normally applied at 0.07–0.15% of the poured weight of iron. The metal temperature and pouring time of the casting must be considered when selecting the tablet weight. A minimum pouring temperature of 1370°C (2500°F) is recommended. It is important that the INOTAB tablet is located where there is continual metal flow during pouring to ensure uniform dissolution and the typical application methods are shown in Fig. 5.5.

Chapter 6
Ductile iron

Production of ductile iron

Ductile iron, also known as spheroidal graphite (s.g.) iron or nodular iron, is made by treating liquid iron of suitable composition with magnesium before casting. This promotes the precipitation of graphite in the form of discrete nodules instead of interconnected flakes (Fig. 2.4). The nodular iron so formed has high ductility, allowing castings to be used in critical applications such as:

Crankshafts, steering knuckles, differential carriers, brake callipers, hubs, brackets, valves, water pipes, pipe fittings and many others.

Ductile iron production now accounts for about 40% of all iron castings and is still growing.

While a number of elements, such as cerium, calcium and lithium are known to develop nodular graphite structures in cast iron; magnesium treatment is always used in practice. The base iron is typically:

TC	Si	Mn	S	P
3.7	2.5	0.3	0.01	0.01

having high carbon equivalent value (CEV) and very low sulphur. Sufficient magnesium is added to the liquid iron to give a residual magnesium content of about 0.04%, the iron is inoculated and cast. The graphite then precipitates in the form of spheroids. It is not easy to add magnesium to liquid iron. Magnesium boils at a low temperature (1090°C), so there is a violent reaction due to the high vapour pressure of Mg at the treatment temperature causing violent agitation of the liquid iron and considerable loss of Mg in vapour form. This gives rise to the familiar brilliant 'magnesium flare' during treatment accompanied by clouds of white magnesium oxide fume. During Mg treatment, oxides and sulphides are formed in the iron, resulting in dross formation on the metal surface, this dross must be removed as completely as possible before casting. It is important to remember that the residual magnesium in the liquid iron after treatment oxidises continuously at the metal surface, causing loss of magnesium which may affect the structure of the graphite spheroids, moreover the dross formed may result in harmful inclusions in the castings.

Several different methods of adding magnesium have been developed, with the aim of giving predictable, high yields. Magnesium reacts with sulphur present in the liquid iron until the residual sulphur is about 0.01%. Until the sulphur is reduced to near this figure, the magnesium has little effect on the graphite formation. In the formation of MgS, 0.1%S requires 0.076%Mg. A measure of the true Mg recovery of the treatment process can be expressed as:

$$\text{Mg recovery \%} = \frac{0.76 \times (\text{S\% in base metal} - \text{S\% residual}) + \text{residual Mg\%}}{\text{Mg\% added}}$$

Mg recovery is lower at high treatment temperatures and is dependent on the particular treatment process used. Magnesium may be added as pure Mg, or as an alloy, usually Mg–ferrosilicon or nickel–magnesium. Other materials include briquettes, called NODULANT, formed from granular mixtures of iron and magnesium and hollow mild steel wire filled with Mg and other materials.

Magnesium content of treatment materials	
Mg–Fe–Si alloy	3–20%
Ni–Mg alloy	5–15%
Mg ingot or wire	>99%
Mg–Fe briquettes	5–15%
Cored wire	40–95%

MgFeSi alloys usually also contain 0.3–1.0% cerium accompanied by other rare earth elements. 0.5–1.0%Ca is also a common addition to the treatment alloy.

Typical analysis of magnesium ferrosilicon nodulariser

Element	*5% MgFeSi*	*10% MgFeSi*
Si %	44–48	44–48
Mg %	5.5–6.6	9.0–10.0
Ca %	0.2–0.6	0.5–1.0
RE %	0.4–0.8	0.4–1.0
Al %	1.2 max	1.2 max

RE (rare earths) contain approximately 50%Ce

Treatment methods include:

Sandwich ladle: the treatment alloy is contained in a recess in the bottom of a rather tall ladle and covered with steel scrap. The method is suitable for use only with treatment alloys containing less than 10% Mg (Fig. 6.1a). *Tundish cover:* this is a development of the treatment ladle in which a specially designed cover for the ladle improves Mg recovery and almost eliminates glare and fume (Fig. 6.1c).

Plunger: the alloy is plunged into the ladle using a refractory plunger bell usually combined with a ladle cover and fume extraction (Fig. 6.1d).

Porous plug: a porous-plug ladle is used to desulphurise the metal with calcium carbide and the treatment alloy is added later while still agitating the metal with the porous plug.

Converter: a special converter-ladle is used, containing Mg metal in a

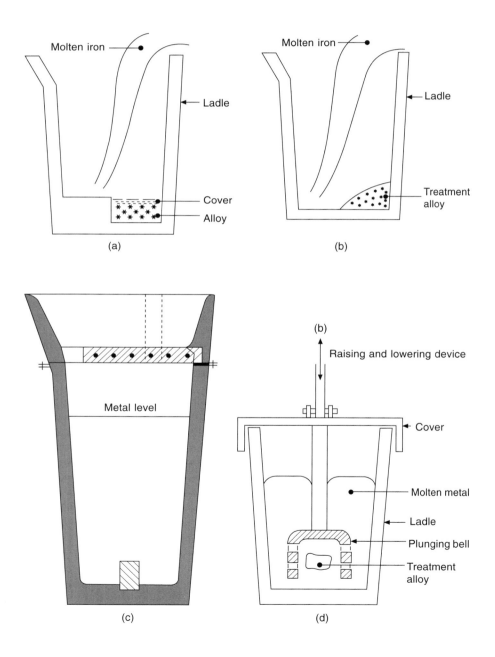

(a)

(b)

(c)

(d)

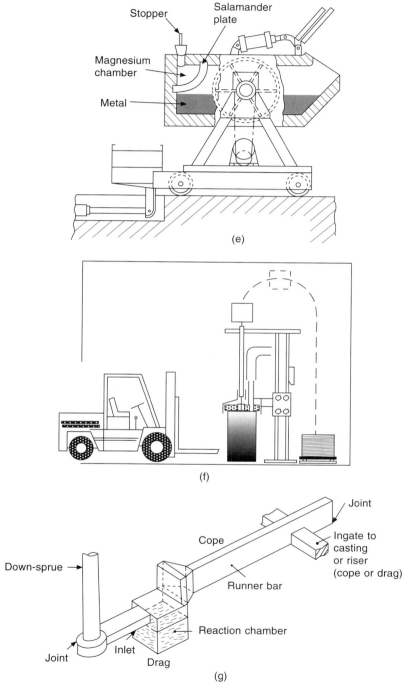

(e)

(f)

(g)

Figure 6.1 *Treatment methods for making ductile iron. (a) Sandwich treatment. (b) Pour-over treatment. (c) Tundish cover ladle. (d) Plunging treatment. (e) GF Fischer converter. (f) IMPREX cored-wire treatment station (g) In-mould system.*

pocket. The ladle is filled with liquid iron, sealed and rotated so that the Mg metal is submerged under the iron (Fig. 6.1e).

Cored wire treatment: wire containing Mg, FeSi, Ca is fed mechanically into liquid metal in a covered treatment ladle at a special station (Fig. 6.1f).

Treatment in the mould (Inmold): MgFeSi alloy is placed in a chamber moulded into the running system, the iron is continuously treated as it flows over the alloy (Fig. 6.1g).

All the methods have advantages and disadvantages; simple treatment methods can only be used with the more costly low-Mg alloys, generally containing high silicon levels which can be a restriction since a low Si base iron must be used. In order to use high Mg alloys and pure Mg, expensive special purpose equipment is needed so the method tends to be used only by large foundries.

A survey on ductile iron practice in nearly 80 US foundries in 1988 (*AFS Trans.* **97**, 1989, p. 79), showed that the biggest change in the previous 10 years was the increase in the use of the tundish ladle, used by over half of the foundries in the survey. The growth had come at the expense of open-ladle, plunging, porous plug and sandwich processes. More recently, cored-wire treatment has been developed and its use is growing.

Melting ductile iron base

While the cupola can be used for the production of ductile iron, the need for high liquid iron temperatures and close composition control has encouraged the use either of duplexing with an induction furnace, or using a coreless induction furnace as prime melter.

In the US survey referred to above, coreless induction furnaces were used by 84% of the smaller foundries (producing less than 200 t/week). Almost all larger foundries duplexed iron from an acid cupola to an induction furnace, with channel furnaces being favourite.

Cupola melting and duplexing

If magnesium treatment with MgFeSi alloy is used, a low Si base iron is needed. The process may be summarised as follows:

Melt in acid cupola, charge foundry returns and steel scrap plus low sulphur pig iron if necessary.

Tap at around 2.8–3.2%C
0.6–1.0%Si
0.08–0.12%S

Desulphurise, using porous plug treatment with calcium carbide, to about 0.10%S, carburise to 3.6–3.8%C.

Transfer to induction furnace, adjust C and Si and temperature to required levels.

Treat with MgFeSi and inoculate.

Cast.

Induction furnace melting

Charge foundry returns, steel scrap, ferrosilicon and carburiser to achieve the desired composition.

If sulphur is below 0.025%, desulphurisation is not necessary, but the higher the sulphur content, the more magnesium must be used so the cost of treatment increases.

Treat with Mg and inoculate.

Cast.

If a converter or cored-wire Mg treatment is used, high silicon base irons are satisfactory. Separate desulphurisation is not necessary since, with these processes it is economical to use pure magnesium as a desulphuriser.

Use of the tundish cover ladle

The most commonly used treatment method, particularly in smaller foundries is the tundish covered treatment ladle. The principle is shown in Fig. 6.1c. The use of a refractory dividing wall to form an alloy pocket in the bottom of the ladle gives improved Mg recovery compared to a pocket recessed in the bottom of the ladle. Treatment batches are usually in the range, 450–1000 kg. Figure 6.2 shows the design of a ladle suitable for the treatment of about 450–500 kg of iron. The diameter of the filling hole is chosen to minimise the generation of fume while allowing the ladle to be filled quickly without excessive temperature loss. It is essential that the MgFeSi alloy is not exposed to the liquid iron until quite late in the filling procedure, so the filling hole is positioned to introduce liquid iron away from the alloy pocket in the ladle bottom. The Mg alloy in the alloy pocket is covered with steel turnings or FeSi pieces of size 25 × 6 mm, then when the level in the ladle reaches the dividing wall, iron flows over and forms a semi-solid mass with the cover material allowing the ladle to be almost filled before the reaction starts, thus ensuring good recovery of Mg.

In order to minimise temperature losses during treatment, the ladle and cover should be separately heated with gas burners before assembly. Immediately before use, the ladle should be filled with base iron from the melting furnace and allowed to soak for a few minutes before returning the iron to the furnace. The prescribed weight of MgFeSi alloy is charged through

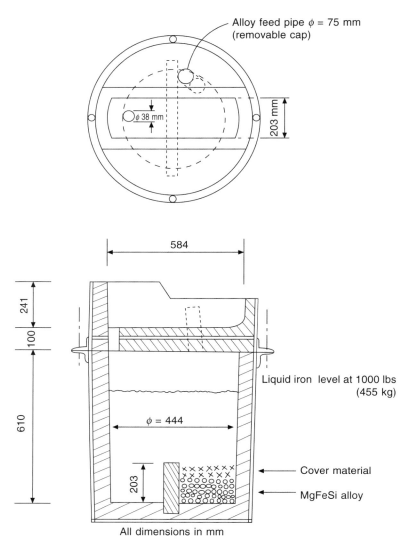

Figure 6.2   *Plan and cross-section of tundish/cover ladle. (From Anderson, J.V. and Benn, D. (1982)* AFS Trans, **90,** *159–162.)*

the alloy charging tube which is plugged after removal of the charging funnel. The treatment alloy may be any of the MgFeSi alloys with Mg in the range 3–6%, additions of 1.5–3.0% are made giving Mg additions of 0.08–0.15%. Tapping time is usually around 40 seconds.

The temperature loss during treatment is around 50°C, so the tapping temperature must be adjusted accordingly, treatment temperatures of around 1530°C are commonly used. After treatment, the tundish cover is removed, the metal transferred to a pouring ladle where inoculation may take place, then it is cast.

Sandwich treatment

A popular method of treatment, frequently used in smaller foundries, is the sandwich method (Fig. 6.1a). This is essentially the same as the tundish-cover method but carried out in an open ladle. The magnesium alloy is placed in a pocket in the bottom of the ladle and covered with steel scrap (2–3% of the metal weight) or a steel plate. The molten metal stream is directed away from the pocket. The pocket must be deep enough to contain all the alloy and the steel scrap which should be of small size to produce a high packing density. The treatment ladle is usually deep, with a height-to-diameter ratio of 1.5–2.0:1, the extra metal depth increases the recovery of the magnesium, which can be as high as 50% when a 5%Mg alloy is used.

NODULANT

NODULANT briquettes are essentially formed of pure magnesium and sponge iron. They contain 10% Mg and minor quantities of calcium, cerium, silicon and carbon. The briquettes weigh between 16 and 20 g and have a density of 4.3–4.5 g/ml, so they are suitable for use in the sandwich or tundish cover technique in just the same way that MgFeSi is used. Magnesium yields are around 40%. The major advantage of using NODULANT is that a negligible amount of silicon is added during the treatment. This permits the use of higher silicon in the base iron enabling all the available ductile iron returns to be used in the charge. By increasing the base silicon, the lining life of the induction furnace is increased by as much as 40%.

Pure magnesium converter process

The Georg Fischer converter (Fig. 6.1e), has a reaction chamber formed by a graphite-clay plate of semi-circular section set into the lining of the converter. Molten iron is charged with the converter in the horizontal position. The reaction chamber is charged with pure Mg lumps and other additives (if required) and sealed with a locking stopper. The converter spout is closed by a pneumatically operated lid. The vessel is then tilted to vertical allowing a limited amount of metal to enter through holes in the chamber and react with the magnesium which starts to vaporise. The vapour pressure in the reaction chamber rises slowing the further entry of liquid iron and allowing controlled treatment of the contents of the converter. Treatment takes 60–90 seconds. The melt is first deoxidised then desulphurised. When the sulphur content has dropped to less than 0.002%, the melt starts to absorb magnesium. Magnesium recovery can be as high as 70%. Base irons with high sulphur contents of 0.2–0.3%S can be used because of the high efficiency of Mg use. The standard converter allows up to 2.5 tonnes of iron to be treated 6–8 times per hour. Larger units with capacity up to 10 tonnes are available. The reaction chamber wall has a limited life of 200–800 treatments. Temperature loss is 22–33°C in a 1-tonne converter but less in larger converters. Since no Si is added during treatment, an unlimited amount of ductile iron returns can be used.

The process is operated under licence from Disa Georg Fischer.

Cored-wire treatment

The equipment and principle of the method is shown in Fig. 6.1f which shows the Foseco IMPREX-Station. The treatment station consists of a coil of hollow mild steel tube filled with Mg metal, a feeding machine, a guide tube and a ladle with a close fitting lid. On entering the molten metal, the sheathing of the wire dissolves releasing the core material below the metal surface. Wires vary in size from 4 mm to 13 mm. The amount of wire needed is dependent on the sulphur content of the base iron, the temperature of the iron and the reactivity of the wire used. Once the treatment parameters have been established, it is a simple matter to calculate the amount of wire required and the treatment time. Feed details can then be programmed into a computer. Typical feed rates of 9 mm wire are 30–50 metres per minute. 1500 kg of metal can be treated in about 2 minutes. Treatment temperature starts at around 1450°C, dropping to around 1410°C at the finish.

In-the-mould treatment

It is possible to carry out the nodularising treatment in the mould by incorporating a specially designed magnesium treatment chamber in the gating system into which the treatment alloy is placed (Fig. 6.1g). The gating method must fill with metal as quickly as possible and must maintain constant flow conditions so that all the metal, from first to last, is equally treated. Since treatment of iron with Mg always produces some MgO and MgS dross, care must also be taken to avoid dross entering the casting. This is not easy to achieve and requires a good deal of experimentation so the method is generally only used by large repetition foundries, which are able to devote considerable time to solving the process problems.

Inhibiting elements

Certain elements which may be present in the base iron have an inhibiting effect on nodule formation, the following elements are known to be harmful:

Aluminium	above 0.13%
Arsenic	above 0.09%
Bismuth	above 0.002%
Lead	above 0.005%
Tin	above 0.04%
Titanium	above 0.04%

Antimony, tellurium and selenium are also harmful. The combined effect of two or more of these elements may be even more harmful. The addition of cerium and other rare earth elements, together with calcium will neutralise many of the harmful effects of inhibiting elements and most MgFeSi nodularising alloys contain 0.3–1.0% Ce and other rare earths. 0.5–1.0% Ca is also commonly present.

Inoculation and fading

Immediately after treatment, the iron must be inoculated. Larger additions of inoculant are needed compared with grey iron and from 0.5–0.75% of a graphitising inoculant such as INOCULIN 25 should be used. Inoculation treatment is not permanent, the effect begins to fade from the time the inoculant is added. As the inoculating effect fades, the number of nodules formed decreases and the tendency to produce chill and mottle increases. In addition the quality of the graphite nodules deteriorates and quasi-flake nodules may occur.

When inoculating ductile iron, the inoculant must be added after the magnesium flare has subsided. A common practice is to tap about half the metal onto the magnesium alloy and wait for the flare to finish before adding the inoculant to the tapping stream as the rest of the metal is tapped. If the metal is transferred from the treatment ladle to a casting ladle, an effective practice is to make a further small addition of inoculant as the metal is poured into the casting ladle. About 0.1–0.2% of inoculant is adequate.

Significant fading occurs within five minutes of inoculation. Because of this problem, late stream or mould inoculation is commonly used in ductile iron production, see p. 68.

Specifications for ductile cast iron

Table 6.1 lists a number of national and international specifications for ductile iron, it is necessary to consult the original specifications for details of the methods of testing and the mandatory values that must be achieved. In recent years, specifications in different countries have been converging so they are now all quite similar. Table 6.2 lists the suggested chemical compositions required to produce castings that meet the specifications in the as-cast state.

Table 6.1 Specifications for ductile (nodular) cast irons

Country Specification	Minimum tensile strength/elongation (N/mm^2/%)								
Europe EN-GJS-CEN 1563:1997	350-22	400-18	400-15	450-10	500-7	600-3	700-2	800-2	900-2
UK* BS2789 1985	350/22	400/18	420/12	450/10	500/7	600/3	700/2	800/2	900/2
USA ASTM A536 1993		60-40-18	65-45-12	70-50-05	80-55-06	80-60-03	100-70-03	120-90-02	
Japan JIS FCD G5502 1995	350-22	400-18	400-15	450-10	500-7	600-3	700-2	800-2	
Inter-national ISO 1083 1987	350-22	400-18	400-15	450-10	500-7	600-3	700-2	800-2	900-2
Hardness Typical HB	<160	130-175	135-180	160-210	170-230	190-270	225-305	245-335	270-360
Typical structures		F	F	F & P	F & P	F & P	P	P or T	TM

Notes: The European CEN 1563 Standard also specifies 350-22-LT and 400-18-LT for low temperatures. 350-22-RT and 400-18-RT for room temperature.

*BS2789 has been withdrawn and replaced by EN1563:1997.

The designations of US Standards e.g. 100-70-03, refers to min. tensile strength(lbf/in^2)-min. proof stress-elongation %.

The structures are: TM, tempered martensite; P or T, pearlite or tempered structure; P & F, pearlite and ferrite; F, ferrite.

This Table is intended only as a guide, refer to the National Standards for details.

Table 6.2 Suggested analyses for as-cast production of ductile iron

Average casting section	Grades 800/2,700/2,600/3			Grade 500/7			Grade 400/12			Grade 400/18		
	TC	Si	Mn(max)	TC	Si	Mn(max)	TC	Si	Mn(max)	TC	Si	Mn(max)
<13 mm	3.6–3.8	2.6–2.8	0.5	3.6–3.8	2.6–2.8	0.3	3.6–3.8	2.6–2.8	0.2	3.6–3.8	2.6–2.8	0.1
13–25	3.5–3.6	2.2–2.5	0.6	3.5–3.6	2.2–2.5	0.35	3.5–3.6	2.2–2.4	0.25	3.5–3.6	2.2–2.4	0.15
25–50	3.5–3.6	2.1–2.3	0.7	3.5–3.6	2.1–2.4	0.4	3.5–3.6	2.2–2.4	0.3	3.5–3.6	2.2–2.4	0.20
50–100	3.4–3.5	1.9–2.1	0.8	3.4–3.5	2.0–2.2	0.5	3.4–3.5	2.0–2.2	0.35	3.4–3.5	1.8–2.0	0.2
>100	3.4–3.5	1.8–2.0	0.8	3.4–3.5	1.8–2.0	0.6	3.4–3.5	1.8–2.0	0.40	3.4–3.5	1.8–2.0	0.25

Notes: For the higher strength grades, 800/2,700,600, additions of 0.5% Cu or 0.1% Sn may be made to encourage pearlite formation.
In all grades: Phosphorus should be less than 0.05%
Chromium should be less than 0.05%
Residual Mg should be 0.03–0.06%

Heat treatment of ductile iron

It is obviously desirable to achieve the required properties in the as-cast form, but this is not always possible because of variations of section thickness etc. Heat treatment of the castings will eliminate carbides in thin sections, produce more consistent matrix structures and for a given structure, the mechanical properties are often improved by heat treatment, especially by normalising. Where tempered martensite structures are needed, heat treatment is essential.

Stress relief

Heat at 50–100°C/h to 600°C (taking care not to exceed 610°C), soak for one hour plus an hour for every 25 mm of section thickness in the thickest section. Cool at 50–100°C/h to 200°C or less. Ensure that the castings are adequately supported in the furnace so that they are not subjected to stress.

Breakdown of carbides

Thin section castings may contain carbides in the as-cast structure, these can be eliminated by soaking the castings at 900–925°C for 3 to 5 hours.

Annealing to produce a ferritic matrix

Castings should be soaked at 900–925°C for 3–5 hours, followed by slow cooling at around 20–35°C/h through the critical temperature (about 800–710°C), then furnace cooled at, say 50–100°C/h to 200°C.

Normalising to produce a pearlitic matrix

Soak the castings above the critical temperature then air cool. Again a soaking temperature of 900–925°C is usually used, to ensure that carbides are broken down, then use forced air cooling to form pearlite. The type of heat treatment furnace available and the size of the load determines the cycle that is possible. It may be necessary to adjust the metal composition with tin or copper to help the formation of fully pearlitic structures.

Hardened and tempered structures

Austenitise at 900–920°C then oil quench. Tempering is usually carried out at 600–650°C.

Austempered ductile iron (ADI)

Austempering is an isothermal heat treatment for producing 'bainitic' structures. It can double the strength of ductile iron while retaining good ductility and toughness. Wear resistance and fatigue properties are excellent so that ADI is comparable with wrought steel.

The ADI heat treatment is a two-stage process, shown in Fig. 6.3. Austenitising is carried out at 815–930°C to fully transform the matrix to austenite. This is done either in a non-oxidising atmosphere furnace or in a high temperature salt bath, temperatures and times are determined by chemical composition, section size and grade of ADI required. 1 to 1.5 hours is usually adequate. Slow initial heating of the casting is desirable to avoid the danger of cracking of complex shapes. The castings are then quenched to the required isothermal heat treatment temperature, usually between 210 and 400°C. This is usually done in a salt bath (Fig. 6.3). The castings are held at temperature for 1–2 hours to complete the transformation of austenite to bainite. The lower temperatures give high hardness, strength and wear resistance, while the higher heat treatment temperatures result in higher ductility and toughness. After the isothermal treatment, the castings are cooled to ambient temperature.

Unalloyed ductile irons may be austempered in sections up to about 8 mm. Thicker section castings require the addition of Mo or Ni to increase the hardenability.

Typical changes in properties due to austempering of an unalloyed iron are:

	As-cast	*Austempered* 1h *at 300°C*	*Austempered* 1h *at 375°C*
Tensile strength (N/mm^2)	475	1465	1105
Elongation (%)	19	1	9
Hardness (HB)	160	450	320

Austempered ductile iron finds applications as a replacement for forged steel components in the agricultural, mining, automotive and general engineering industries; for example: plough tips, digger teeth, spring brackets, rear axle brackets, gears etc. ADI production is growing but its use is limited to some extent by the lack of suitable heat-treatment facilities.

The European/British Specification BS EN 1564:1997 defines four grades of ADI, Table 6.3.

Table 6.3 European grades of ADI

Material designation	*Tensile strength* (MPa)	*0.2% PS* (MPa)	*Elongation* (%)	*Hardness* (HB)
EN-GJS-800-8	800	500	8	260–320
EN-GJS-1000-5	1000	700	5	300–360
EN-GJS-1200-2	1200	850	2	340–440
EN-GJS-1400-1	1400	1100	1	380–480

Mechanical properties are measured on test pieces machined from separately cast test pieces.

In North America, the ASTM has defined five standard grades of ADI, Table 6.4.

Table 6.4   The five ASTM standard ADI grades (ASTM A897M-90)

Grade	Tensile* strength (MPa)	Yield* strength (MPa)	Elongation* (%)	Impact energy* (Joules)	Typical hardness (BHN)
1	850	550	10	100	269–321
2	1050	700	7	80	302–363
3	1200	850	4	60	341–444
4	1400	1100	1	35	388–477
5	1600	1300	N/A	N/A	444–555

*Minimum values

Casting ductile iron

Ductile iron differs from grey iron in its casting characteristics in two important respects. Unlike grey iron, ductile iron is a dross-forming alloy. The residual magnesium which is needed to ensure nodular graphite formation rapidly oxidises whenever the liquid metal is exposed to air; in the ladle, during metal transfer and in the mould. A magnesium silicate dross is formed which may give rise to defects at or just below the casting surface, usually on the upper surfaces of the castings. For this reason, it is common practice to filter ductile iron castings through ceramic filters, see Chapter 18.

The other major difference compared with grey iron, is the need to feed ductile iron castings to ensure freedom from shrinkage defects. Ductile irons always have a high carbon equivalent so the volume of graphite precipitated during solidification should ensure completely sound sections. However, the expansion resulting from the graphite precipitation results in large pressures being exerted on the mould walls, much higher than those found with grey iron (figures of 1000–1500 kPa or 145–217 lbf/in^2 have been measured compared with 170–200 kPa or 24–29 lbf/in^2 in grey iron). Only the strongest moulds, such as sodium silicate bonded moulds or well-vibrated lost foam moulds, will resist such pressures so the use of feeders is common when ductile iron castings are made.

Compacted graphite irons

Compacted graphite (CG) irons are a range of cast irons having mechanical properties intermediate between those of grey and ductile irons. Under the microscope the graphite appears as short, thick flakes with rounded ends, readily distinguished from true flake graphite (Fig. 6.4). Compacted graphite

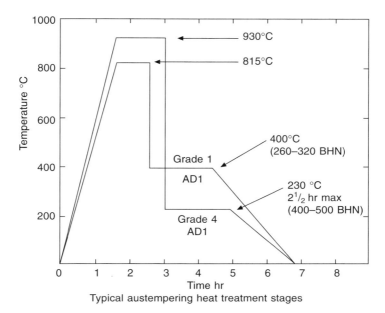

Typical austempering heat treatment stages

Figure 6.3 *Typical austempering heat-treatment stages. (R.D. Forrest, 13th DISA/ GF Licensee Conference 1997. Courtesy Rio Tinto Iron and Titanium GmbH.)*

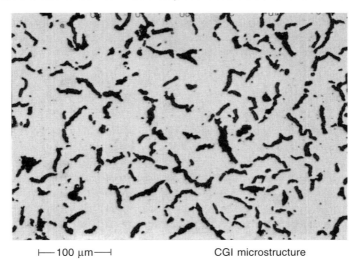

⊢—100 μm—⊣   CGI microstructure

Figure 6.4 *Structure of compacted graphite. (Courtesy SinterCast.)*

is interconnected in a branched structure, and is more like flakes than the completely isolated nodules in ductile iron. Compacted graphite is not the same as vermicular graphite which can occur in irons of very low sulphur.

Compacted graphite iron has good casting characteristics, due to its high fluidity and low solidification shrinkage. The tensile, yield and fatigue strengths of CG iron are 1.5–2 times that of grey iron, approaching ductile iron. Thermal conductivity is comparable with grey iron and machinability is intermediate between grey and ductile irons of similar matrix structures. This combination of properties makes CG iron suitable for upgrading of castings traditionally produced of grey iron.

Production of compacted graphite iron

There are several methods by which compacted graphite may be produced:

Cerium additions
Magnesium additions
Nitrogen additions
Magnesium plus titanium additions

One method involves the joint addition of a nodularising and a de-nodularising agent, usually magnesium and titanium often with a small cerium addition as well. Special treatment alloys have been developed for the production of CG iron.

Mg	Ti	Ce	Ca	Al	Si	Fe
4.5–5.5	8.0–10.0	0.3–0.4	1.0 max	1.5 max	50–54	balance

As with ductile iron, it is desirable to start with a low sulphur iron, below 0.02%. The base iron for treatment is best melted in an induction furnace and should have composition in the range:

CE	C	Si	Mn	S	P
3.7–4.5	3.1–3.9	1.7–2.9	0.1–0.6	0.035 max	0.06 max

A sandwich technique may be used with addition rates between 0.6–1.6% depending on the foundry conditions. The treatment temperature should be above 1350°C to avoid the formation of a fully nodular structure. Inoculation of the treated iron is necessary and additions of 0.2–0.5% ferrosilicon are common. The matrix structure may be made ferritic or pearlitic as with ductile iron. Titanium-containing alloys can cause problems due to build-up of Ti in foundry returns leading to impaired machinability.

The use of compacted graphite iron was limited for many years by the difficulty foundries experienced in controlling the process within the narrow range of CGI stability so it was considered an unreliable material.

A Swedish company, SinterCast, has developed a method of process control for CGI which allows CGI to be made reliably. The process can be licensed by foundries. Figure 6.5 represents the transition of graphite morphology from flake to compacted and ultimately to spheroidal graphite as a function of Mg content. This suggests that a range of 0.005–0.010% total Mg will cause formation of compacted graphite. This curve does not fully account for reactions between Mg and dissolved elements in the iron, particularly oxygen and sulphur. Figure 6.6 illustrates by deep etched scanning

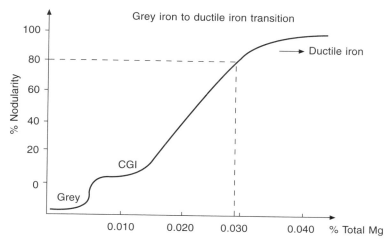

Figure 6.5 *The effect of total Mg content on the transformation of graphite morphology from flake to compacted and ultimately to spheroidal graphite. (Courtesy SinterCast.)*

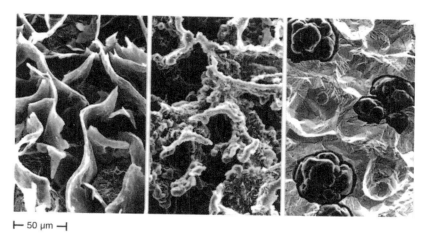

Figure 6.6 *The transition of graphite morphology from flake to compacted to spheroidal shown by deep etched scanning electron micrographs. (Courtesy SinterCast.)*

electron micrographs, the transition of graphite morphology from flake to compacted to spheroidal. The SinterCast Process involves taking a sample of liquid iron from the ladle in a patented sampling cup and carrying out a thermal analysis using two thermocouples, one in the centre of the cup and the other adjacent to the wall. The two cooling curves allow the degree of modification and inoculation required to be determined. Independent additions of cored wire containing Mg and inoculant are then made to the ladle. Ladle to ladle variations in the oxygen and sulphur content of the iron are thus allowed for and reliable compacted graphite iron is made.

Foundry properties of compacted graphite iron

The fluidity is governed by carbon equivalent (CE) and temperature and is similar to grey or ductile irons of the same CE. Because CG irons are stronger than grey irons, a higher CE can be used to obtain the same strength, this allows greater fluidity and easier running of thin sections. CG iron is dross-forming, just as ductile iron, and filtration of castings is desirable. CG irons are more prone to chill than grey irons but less likely to chill than ductile iron. Good inoculation is necessary. Fading occurs, but to a smaller extent than in ductile iron, but excessive delays between treatment, inoculation and casting should be avoided.

There is some disagreement about the level of feeding required for CG iron. There is less tendency for mould wall movement than with ductile iron, nevertheless some feeding appears to be desirable.

Applications of compacted graphite irons

There has been great interest in its use for automotive castings such as diesel cylinder blocks and heads, hydraulic components, exhaust manifolds, brake drums, brake discs, flywheels etc. The lack of consistency of properties held back the wide scale application of CG iron but with the greater control now possible its use is expected to develop.

Properties of compacted graphite irons

Table 6.5   Comparison of CG iron properties with grey and ductile irons

Property	Grey irons	CG irons	Ductile irons
Tensile strength			
ton/in^2	11–20	20–38	26–45
lb/in^2	25–45 000	45–85 000	60–100 000
kg/mm^2	16–32	30–60	40–70
N/mm^2	160–320	300–600	400–700
Elongation (%)	nil	3–6	6–25
Modulus (lb/in^2)	14–16×10^6	20–23×10^6	25–27×10^6
(GN/m^2)	96–110	140–160	170–190
Charpy impact			
Joules, 25°C	nil	3–7	17
Fatigue limit			
un-notched (ton/in^2)	7–8	15–20	12–18
(N/mm^2)	108–123	230–310	185–280
Machinability	very good	very good	good
Corrosion resistance	moderate	intermediate	good

Table 6.6 Specifications for compacted graphite iron: ASTM A842-85 (reapproved 1991) compacted graphite iron

	Grade 250[a]	300	350	400	450[b]
Tensile strength (MPa)	250	300	350	400	450
Yield strength (MPa)	175	210	245	280	315
Elongation (%)	3.0	1.5	1.0	1.0	1.0
Hardness (HB)	179 max	143–207	163–229	197–225	207–269

[a]ferritic grade
[b]pearlitic
Hardness not mandatory

Chapter 7
Malleable cast iron

Introduction

Malleable irons are cast white, that is, their as-cast structure consists of metastable carbide in a pearlitic matrix. The castings must then be annealed to convert the brittle carbide structure and develop a structure of roughly spherical graphite aggregates in a matrix which can be either ferritic or pearlitic, depending on composition and heat treatment.

There are two types of malleable iron, blackheart and whiteheart. Malleable iron has a long history, whiteheart iron having been developed in 1722 by the French metallurgist, Réaumur, while blackheart iron was developed in the USA in 1820. Malleable iron was widely used for automotive and agricultural components, pipe fittings, valves etc. but since the development of spheroidal graphite ductile iron its use has declined, due to the high cost of the annealing treatment which requires expensive furnace equipment. Malleable iron is still widely used for small pipe fittings, electrical fittings and builders hardware, particularly for thin section castings and castings which are subsequently galvanised.

Whiteheart malleable

In the whiteheart process, the white, as-cast iron is decarburised during annealing leaving a structure of iron carbide in a metallic matrix. When fractured, the appearance is whitish, giving rise to the name, 'whiteheart'. Decarburisation is only possible in thin sections; in heavier sections, some conversion of carbide to graphite nodules occurs so that the annealed casting has a white rim with a core having different structure and mechanical properties. This limits the applications to which it can be put. Whiteheart can be melted in a cupola and is a low cost material which still finds applications in small, thin section castings.

Composition of whiteheart malleable

Typical compositions are

	Before annealing	After annealing
Total carbon	3.0–3.7%	0.5–2.0%
Silicon	0.4–0.8	0.4–0.7
Manganese	0.1–0.4	0.1–0.4
Sulphur	0.3 max	0.3 max
Phosphorus	0.1 max	0.1 max

Annealing is a combined decarburisation and graphitisation process performed in an oxidising atmosphere. Originally it was done by packing castings into iron ore mixtures but now it is carried out in continuous, atmosphere controlled furnaces at about 1070°C. Small castings may be fully decarburised and are referred to as weldable malleable irons. Table 7.1 lists the European specifications for whiteheart malleable iron. The USA has no equivalent standard.

Table 7.1 Specifications for whiteheart malleable cast irons

Grade	Test bar dia. (mm)	Tensile strength (min) (N/mm^2)	Elongation (%)	0.2% Proof strength (min) (N/mm^2)	HB typical
EN-GJMW-300-4	12	350	4	–	230
EN-GJMW-360-12	12	360	12	190	200
EN-GJMW-400-5	12	400	5	220	220
EN-GJMW-450-7	12	450	7	260	260
EN-GJMW-550-4	12	550	4	340	340
International ISO 5922–1981 Whiteheart					
W 35-04	9	340	5	–	
	12	350	4	–	230 max
	15	360	3	–	
W 38-12	9	320	15	170	
	12	380	12	200	200 max
	15	400	8	210	
W 40-05	9	360	8	200	
	12	400	5	220	220 max
	15	420	4	230	
W 45-07	9	400	10	230	
	12	450	7	260	150 max
	15	480	4	280	

The European Standard CEN 1562:1997 has superseded the former national standards, for example:
France: NF A32-701(1982); Germany: DIN 1692 (1982); UK: BS6681:1986.
Notes: It is advisable to consult the original standards for details of the mandatory values, methods of testing etc.
USA has no standard for whiteheart malleable iron.

Blackheart malleable iron

The iron is typically melted in a cupola and duplexed into an electric furnace where temperature and composition are adjusted. The cupola metal has the composition:

C	Si	Mn	S	P
2.5–2.6	1.1	0.2–0.3	0.2	0.1

The final composition in the electric furnace is:

C	Si	Mn	S	P	Cr
2.4–2.6	1.3–1.45	0.4–0.55	0.2 max	0.1 max	0.05 max

Castings are poured at around 1450°C. Sometimes small additions (around 0.01%) of bismuth are added in the ladle to ensure fully white as-cast structures. White irons contract on solidification so to ensure freedom from shrinkage, the castings must be fed, see Chapter 19. As-cast malleable iron is extremely brittle, allowing feeders and running systems to be broken easily from the castings. Castings which have abrupt changes of section develop internal stresses on cooling which may be enough to cause cracking of the castings after shakeout. Slow cooling in the mould may be needed to avoid this happening.

The castings are annealed to develop the required graphite clusters (Fig. 2.3). A typical cycle is about 48 hours long (Fig. 7.1) it may be carried out in batch-type or continuous furnaces in a controlled atmosphere to avoid oxidation of the castings. The rate of cooling in the final section of the heat

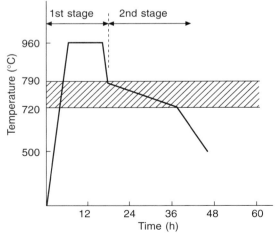

Figure 7.1 *Typical heat treatment cycle for a short cycle blackheart malleable iron. (From Elliott, R. (1988) Cast Iron Technology, Butterworth-Heinemann, reproduced by permission of the publisher.)*

treatment determines the matrix structure of the castings which can be ferritic or pearlitic according to the physical properties required.

Specifications for malleable cast irons

Table 7.2 lists the European and International specifications for blackheart malleable cast irons. The European Standard CEN 1562:1997 has superseded the former national standards. Table 7.3 lists the US ASTM specifications.

Table 7.2   European and international specifications for blackheart malleable cast irons

Grade	Test bar dia. (mm)	Tensile strength (min.) (N/mm²)	Elongation (%)	0.2% Proof stress (min.) (N/mm²)	HB typical
EN-GJMB-300-6	12 or 15	300	6	–	150 max
EN-GJMB-350-10	12 or 15	350	10	200	150 max
EN-GJMB-450-6	12 or 15	450	6	270	150–200
EN-GJMB-500-5	12 or 15	500	5	300	165–215
EN-GJMB-550-4	12 or 15	550	4	340	180–230
EN-GJMB-600-3	12 or 15	600	3	390	195–245
EN-GJMB-650-2	12 or 15	650	2	430	210–260
EN-GJMB-700-2	12 or 15	700	2	530	240–290
EN-GJMB-800-1	12 or 15	800	1	600	270–320
International Standard ISO 5922-1981 Blackheart					
B30-06	12 or 15	300	6		
B32-12	12 or 15	320	12	190	150 max
B35-10	12 or 15	350	10	200	
Pearlitic					
P45-06	12 or 15	450	6	270	150–200
P50-05	12 or 15	500	5	300	160–220
P55-04	12 or 15	550	4	340	180–230
P60-03	12 or 15	600	3	390	200–250
P65-02	12 or 15	650	2	430	210–260
P70-02	12 or 15	700	2	530	240–290
P80-01	12 or 15	800	1	600	270–310

The European Standard CEN 1562:1997 has superseded the former national standards, for example:
France: NF A32-702(1982); Germany: DIN 1692 (1982); UK: BS6681:1986.
Notes: It is advisable to consult the original standards for details of the mandatory values, methods of testing etc.

Table 7.3 US specifications for blackheart malleable irons

Specification	Grade	Test bar diameter (mm)	Tensile strength (N/mm^2)	Elong. (%)	Yield stress (N/mm^2)	Hardness (HB)
ASTM A 47-90 Ferritic	220M10 32510	15.9	340 50000 psi	10	220 32000 psi	156 max
ASTM A220-88 Pearlitic	280M10 40010	15.9	400 60000 psi	10	280 40000 psi	149–197
	310M8 45008	15.9	450 65000 psi	8	310 45000	156–197
	310M6 45006	15.9	450 65000 psi	6	310 45000	156–207
	340M5 50005	15.9	480 70000 psi	5	340 50000 psi	179–229
	410M4 60004	15.9	550 80000 psi	4	410 60000 psi	197–241
	480M3 70003	15.9	590 85000	3	480 70000 psi	217–269
	550M2 80002	15.9	650 95000 psi	2	550 80000 psi	241–285
	620M1 90001	15.9	720 105000 psi	1	620 90000 psi	269–321
ASTM A602-94 Automotive malleable castings	M3210 M4504 M5003 M5503 M7002 M8501			(10) (4) (3) (3) (2) (1)	(32000 psi) (45000 psi) (50000 psi) (55000 psi) (70000 psi) (85000 psi)	156 max[1] 163–217[2] 187–241[2] 187–241[3] 229–269[3] 269–302[3]

ASTM A47-90 (reapproved 1995) covers ferritic malleable irons, a metric version was reapproved in 1996.
ASTM A220-88 (reapproved 1993 including a metric version) covers pearlitic malleable irons.
ASTM A602-94 covers automotive malleable castings.
Note: ASTM A602-94 specifies only hardness and heat treatment. Annealed[1] Air-quenched[2] and tempered[3] Liquid quenched and tempered.

Chapter 8

Special purpose cast irons

Heat resisting alloys

Unalloyed cast iron shows only slight scaling and no growth at temperatures up to 350°C for times of up to 10 years. At 450°C scaling and growth will occur in less than one year in most grey, ductile and malleable cast irons. To achieve greater resistance to heat, special purpose irons must be used.

Temperatures up to 600°C

Unalloyed grey irons can be used up to about 600°C. The presence of phosphorus up to 1.0% improves the scaling resistance and can be used in applications such as fire bars where mechanical strength is not a major consideration.

Temperatures up to 700°C

The addition of about 0.5% Cr improves the oxidation resistance of grey irons and is also used for fire bars. Automotive exhaust manifolds and turbo-charger casings are subjected to increasingly high temperatures in modern engines, unalloyed cast iron can be used at temperatures up to about 600°C, but for higher temperature service it is necessary to alloy the iron with Cr and Mo, e.g.

Low alloy grey iron

C	Si	Mn	S	P	Cr	Mo
3.0–3.4	1.6–2.8	0.4–1.0	<0.12	0.1–0.4	0.5–2.0	0.3–0.5

Temperatures up to 750°C

Ductile (spheroidal graphite) irons with silicon in the range 4–6% have good resistance to scaling and growth and good dimensional stability up to about 750°C. Molybdenum up to 0.5% may be added to improve the high temperature creep resistance. They are used for exhaust manifolds and turbo-charger casings and are better able to resist thermal stress than grey iron castings. Typical compositions are:

High silicon ductile iron

C	Si	Mn	P	Mo
3.2	4.0	0.3 max	0.05 max	0.5
2.9	4.0	0.3 max	0.05 max	0.5
2.6	6.0	0.3 max	0.05 max	0.5

Temperatures up to 850°C

For heat resistance up to 850°C, nickel alloy cast irons, known as Ni-resist (a trade name of the International Nickel Company) are used. Nickel in combination with Mn and Cu produces a stable austenitic matrix and Cr in combination with Ni forms an effective oxidation resistant scale. Most countries have similar standard specifications for austenitic cast irons, based on the original Ni-Resist specifications. Each country designates the alloys differently. Table 8.1 lists the designations of the flake graphite austenitic irons. Table 8.2 lists the specifications of austenitic flake graphite irons.

Spheroidal graphite Ni-resist can also be made. Table 8.3 lists the international designations. Table 8.4 lists the British specifications.

Temperatures up to 1000°C

High Cr white irons have excellent resistance to oxidation at temperatures up to 1000°C and can be used in applications where they are not subjected to impact loading. They are used for furnace parts, sinter pallets, recuperator tubes etc. Typical analyses are:

C	Si	Mn	S	P	Cr
2.0	1.4	1.0	0.1	0.1 max	14.0
2.8	1.8	1.6			17.0
1.0	1.0	0.7			30.0
1.3	1.3	1.0			33.0

Ingot moulds

Products such as ingot moulds, slag ladles, pig-moulds etc. are subjected in use to severe thermal cycling which induces distortion and crazing. High carbon iron is used for such applications, typically: *(Continued on p. 101)*

Table 8.1 Designations of austenitic cast irons with flake graphite

Trade name	ISO 2892:1973	France NF A 32-301 1972	Germany DIN 1694 1981	UK BS 3468:1986	USA ASTM A436 1984 (reapproved 1994)
–	L-NiMn 13 7	L-NM 13 7	GGL-NiMn 13 7	–	–
Ni-Resist 1	L-NiCuCr 15 6 2	L-NUC 15 6 2	GGL-NiCuCr 15 6 2	F1	Type 1
Ni-Resist 1b	L-NiCuCr 15 6 3	L-NUC 15 6 3	GGL-NiCuCr 15 6 3	–	Type 1b
Ni-Resist 2	L-NiCr 20 2	L-NC 20 2	GGL-NiCr 20 2	F2	Type 2
Ni-Resist 2b	L-NiCr 20 3	L-NC 20 3	GGL-NiCr 20 3	–	Type 2b
Nicrosilal	L-NiSiCr 20 5 3	L-NSC 20 5 3	GGL-NiSiCr 20 5 3	–	–
Ni-Resist 3	L-NiCr 30 3	L-NC 30 3	GGL-NiCr 30 3	F3	Type 3
Ni-Resist 4	L-NiSiCr 30 5 5	L-NSC 30 5 5	GGL-NiSiCr 30 5 5	–	Type 4
–	L-Ni 35	L-N 35	–	–	Type 5
–	–	–	–	–	Type 6

Table 8.2 ISO and UK specifications for austenitic flake graphite irons

ISO grade	UK	C max	Si	Mn	Ni	Cr	Cu	Tensile strength (kgf/mm²)	(N/mm²)	(ton/in²)
L-NiMn 13 7	–	3.0	1.5–3.0	6.0–7.0	12.0–14.0	0.2 max	0.5 max	14.3	140	9.1
L-NiCuCr 15 6 2	F1	3.0	1.0–2.8	0.5–1.5	13.5–17.5	1.0–2.5	5.5–7.5	17.3	170	11.0
L-NiCuCr 15 6 3	–	3.0	1.0–2.8	0.5–1.5	13.5–17.5	2.5–3.5	5.5–7.5	19.4	190	12.3
L-NiCr 20 2	F2	3.0	1.0–2.8	0.5–1.5	18.0–22.0	1.0–2.5	0.5 max	17.3	170	11.0
L-NiCr 20 3	–	3.0	1.0–2.8	0.5–1.5	18.0–22.0	2.5–3.5	0.5 max	19.4	190	12.3
L-NiSiCr 20 5 3	–	2.5	4.5–5.5	0.5–1.5	18.0–22.0	1.5–4.5	0.5 max	19.4	190	12.3
L-NiCr 30 3	F3	2.5	1.0–2.0	0.5–1.5	28.0–32.0	2.5–3.5	0.5 max	19.4	190	12.3
L-NiSiCr 30 5 5	–	2.5	5.0–6.0	0.5–1.5	29.0–32.0	4.5–5.5	0.5 max	17.3	170	11.0
L-Ni 35	–	2.4	1.0–2.0	0.5–1.5	34.0–36.0	0.2 max	0.5 max	12.2	120	7.8

Note: Refer to the standards for further details, the above figures are intended only as a guide.

Table 8.3 Designations of heat-resisting austenitic spheroidal graphite irons

International trade name	ISO 2892	UK BS 3468:1986	France NF A 32-301	Germany DIN 1694	US ASTM A439-83 (reapproved 1994)
–	S-NiMn 13 7	S6	S-NM 13 7	GGG-NiMn 13 7	–
Ni-Resist D-2	S-NiCr 20 2	S2	S-NC 20 2	GGG-NiCr 20 2	D-2
Ni-Resist D-2W	–	S2W	–	GGG-NiCrNb 20 2	–
Ni-Resist D-2B	S-NiCr 20 3	S2B	S-NC 20 3	GGG-NiCr 20 3	D-2B
Nicrosilal Spheronic	S-NiSiCr 20 5 2	–	S-NSC 20 5 2	GGG-NiSiCr 20 5 2	–
Ni-Resist D-2C	S-Ni 22	S2C	S-N 22	GGG-Ni 22	D-2C
Ni-Resist D-2M	S-NiMn 23 4	S2M	S-NM 23 4	GGG-NiMn 23 4	–
Ni-Resist D-3A	S-NiCr 30 1	S3	S-NC 30 1	GGG-NiMn 30 1	D-3A
Ni-Resist D-3	S-NiCr 30 3	–	S-NC 30 3	GGG-NiCr 30 3	D-3
Ni-Resist D-4A	–	–	–	GGG-NiSiCr 30 5 2	–
Ni-Resist D-4	S-NiSiCr 30 5 5	–	S-NSC 30 5 5	GGG-NiSiCr 30 5 5	D-4
Ni-Resist D-5	S-Ni 35	–	S-N 35	GGG-Ni 35	D-5
Ni-Resist D-5B	S-NiCr 35 3	–	S-NC 35 3	GGG-NiCr 35 3	D-5B
Ni-Resist D-5S	–	S5S	–	GGG-NiSiCr 35 3 2	D-5S

Note: For detailed specifications refer to the national and international standards.

Table 8.4   British Specification BS 3468(1986) Heat-resisting austenitic spheroidal graphite irons

Grade	C	Si	Mn	Ni	Cu	Cr	Nb	P	Mg	TS(N/mm^2)	Elong. (%)	HB	Impact J, min.
S2	3.0	1.5–2.8	0.5–1.5	18–22	0.5	1.5–2.5	–	0.08	–	370–490	7–20	140–230	–
S2B	3.0	1.5–2.8	0.5–1.5	18–22	0.5	2.5–3.5	–	0.08	–	370–490	7–20	140–230	4
S2C	3.0	1.5–2.8	1.5–2.5	21–24	0.5	0.5	–	0.08	–	370 min	20 min	140–220	20
S2M	3.0	1.5–2.5	4.0–4.5	21–24	0.5	0.2	–	0.08	–	420 min	25 min	160–250	15
S2W	3.0	1.5–2.2	0.5–1.5	18–22	0.5	1.5–2.2	0.12–0.2	0.05	0.06	370–490	7–20	140–200	–
S3	2.5	1.5–2.8	0.5–1.5	28–32	0.5	2.5–3.5	–	0.08	–	370–490	7–20	140–230	–
S5S	2.2	4.8–5.4	1.0	34–36	0.5	1.5–2.5	–	0.08	–	370–470	7–20	130–180	–
S6	3.0	1.5–2.5	6–7	12–14	0.5	0.2	–	0.08	–	390 min	15 min	120–200	–

Note: Refer to British Specification BS 3468(1986) for further details, the above figures are intended only as a guide.

C	Si	Mn	S	P
3.6–3.9	1.4–1.8	0.7–1.0	as low as possible	

The iron is of eutectic composition or slightly hyper-eutectic, since this is found to give the maximum mould life. Compacted graphite iron is also used for this application (see p. 84).

Corrosion resistant cast irons

Unalloyed cast irons exhibit reasonable corrosion resistance, particularly to alkali environments, but special alloys have been developed for use in corrosive conditions. These include, high nickel (Ni-Resist), high silicon and high chromium alloys. Table 8.5 summarises the corrosion resistance of various alloy irons in different media.

Table 8.5 Corrosion resistance of cast irons

	Unalloyed grey, ductile and malleable	Ni-Resist See Table 8.1	High silicon 10–16% Si	High chromium 30% Cr, 1–2% C
Atmospheric including coastal	B	A	A	A
Water soft	B	A	A	A
hard	A	A	A	A
sea	C	A	A	A
Dilute acid	C	B	A	A[1]
Concentrated acid	C[2]	C	A[3]	C
Dilute alkali	A	A	C	A
Concentrated alkali	B	A	C	C

Notes: A – Good resistance   1 Unalloyed iron resists concentrated sulphuric acid quite well

B – Medium resistance   2 High silicon irons do not resist hydrofluoric acid

C – Poor resistance   3 High chromium irons are less resistant to hydrochloric acid

Ni-Resist alloy irons are used for the manufacture of pumps for the handling of sulphuric acid and alkalis, for sea water pumps and for handling sour crude oils in the petroleum industry.

High silicon cast irons, 10–16% Si, are suitable for use in chemical plant and for cathodic protection anodes. The grades shown in Table 8.6 are used.

High silicon irons are low strength, brittle materials which are susceptible to cracking due to stresses induced by cooling. Moulds and cores used for their manufacture should have good breakdown properties. Electric furnace melting is necessary because of their low carbon content. The alloys are

Table 8.6   High silicon cast irons

Grade	C (max)	Si	Mn (max)	S (max)	P (max)	Cr
Si 10	1.2	10.00–12.00	0.5	0.1	0.25	–
Si 14	1.0	14.25–15.25	0.5	0.1	0.25	–
SiCr 14 4	1.4	14.25–15.25	0.5	0.1	0.25	4.0–5.0
Si 16	0.8	16.00–18.00	0.5	0.1	0.25	–

Grade   Si 14 is used for general applications
             Si 10 has higher tensile strength but lower corrosion resistance
             Si 16 has high corrosion resistance but lower strength
             Si Cr 14 4 is used for the manufacture of cathodic protection anodes
For further details, refer to BS 1591.

prone to unsoundness due to hydrogen gas and care must be taken during melting to avoid contact with hydrogen, by the avoidance of moisture on charge materials and refractories and maintaining low levels of Al in the iron.

Heat treatment is advisable to eliminate internal stresses, the castings should be stripped from the moulds as hot as possible and charged into a furnace at about 600°C then heated to 750–850°C for 2–8 hours, depending on metal section. They should then be slowly cooled to below 300°C before removing from the furnace.

High silicon irons have high hardness and are difficult to machine; grinding is preferred.

Wear resistant cast irons

Wear resistant alloy irons are an important class of cast irons, they find wide application for the manufacture of mineral crushing plant, e.g hammers for rock crushers, grinding balls, liners for crushing mills etc. They are also used in shot blasting machines for impellers, liners etc. Three types of cast iron are used:

Unalloyed and low alloy grades of white iron having a structure of massive carbides in a pearlitic matrix. These alloys are extremely brittle and have been largely superseded by tougher, alloyed white irons.
Ni-hard irons containing Ni to increase hardenability by ensuring that austenite transforms to martensite after heat treatment. They also contain Cr to increase the hardness of the carbide.
High Cr-Mo white irons combining abrasion resistance with toughness. The Mo increases hardenability allowing heavy section castings to be made with a martensitic structure, either as cast or heat treated.

Low alloy cast irons

These include irons having the compositions:

BS 4844:1986	C	Si	Mn	Cr
1A,1B,1C	2.4–3.4	0.5–1.5	0.2–0.8	2.0 max

Grade	Hardness (min) HB Thickness < 50 mm	Thickness > 50 mm
1A	400	350
1B	400	350
1C	250	200

Such irons are extremely brittle and have been largely superseded by alloyed irons.

Ni-hard irons

The first alloyed irons to be developed were the Ni-hards (a trade name of the International Nickel Company). They contain Ni to increase hardenability, ensuring that austenite transforms to martensite after heat treatment. The Cr content increases the hardness of the eutectic carbide. The choice of alloy composition depends on casting thickness and intended use. For maximum abrasion resistance, the C content should be at the top of the range, 3.2–3.6%, but for impact resistance it is usual to use lower carbon contents, 2.7–3.2%. Table 8.7 lists a number of national specifications for alloy irons of the Ni-hard type. Note that the correspondence between specifications is not exact and the standards themselves must be referred to for details.

Ni-hards are usually stress relieved at 200–230°C for at least 4 hours to improve strength and impact resistance.

High Cr-Mo irons

The addition of Cr in the range 12–28%, together with Ni and Mo forms a range of alloys which combines abrasion resistance with toughness suitable for large size equipment in the mining, coal and mineral processing industries. Eutectic carbides of the form, M_7C_3 are formed and the matrix can be austenitic, martensitic or pearlitic according to the application. Some components may be cast pearlitic to aid machining, then heat-treated to give an abrasion resistant martensitic structure. Tables 8.8 and 8.9 list specifications of this type.

Production of high chromium irons

An induction furnace is normally used to achieve the required temperature.

Table 8.7 European specifications for abrasion resistant white irons

Trade name	Country	Specification	Grade	C	Si	Mn	Ni	Cr	Cu	Mo	Pmax	Smax	Hardness HB min As-cast	Hardened	Annealed
Ni-hard 1	France	NF A 32-401 1980	FB Ni4 Cr2 HC	3.2–3.6	0.2–0.8	0.3–0.7	3.0–5.5[1]	1.5–2.5	–	0.0–1.0	–			500–700	–
	Germany	DIN 1695 1981	G-X330 NiCr4 2	3.0–3.6	0.2–0.8	0.3–0.7	3.3–5.0	1.4–2.4	–	0.5			450(430)[2]	550(500)	–
	UK	BS 4844:1986	2B	3.2–3.6	0.3–0.8	0.2–0.8	3.0–5.5	1.5–2.5	–	0.5 max	0.3	0.15		550[3]	
Ni-hard 2	France		FB Ni4 Cr2 BC	2.7–3.2	0.2–0.8	0.3–0.7	3.0–5.5[1]	1.5–2.5	–	0.0–1.0	–	–		450–650	
	Germany		G-X260 NiCr4 2	2.6–2.9	0.2–0.8	0.3–0.7	3.3–5.0	1.4–2.4	–				450(430)[2]	520(480)	–
	UK		2A	2.7–3.2	0.3–0.8	0.2–0.8	3.0–5.5	1.5–2.5	–	0.5 max	0.3	0.15		500[3]	
Ni-hard 3	France		FB A	2.7–3.9	0.4–1.5	0.2–0.8	03–3.0[4]	0.2–2.0[4]	0.3–2.0[4]	0.1–1.0[4]	–	–		400–600	
	Germany		–												
	UK		–												
Ni-hard 4	France		FB Cr9 Ni5	2.5–3.6	1.5–2.2	0.3–0.7	4.0–6.0	5.0–11.0	–	0.5 max				550–750	
	Germany		G-X 300 CrNiSi 9 5 2	2.5–3.5	1.5–2.2	0.3–0.7	8.0–10.0	4.5–6.5	–	0.5			450(430)	600(535)	
	UK		2C	2.4–2.8	1.5–2.2	0.2–0.8	4.0–6.0	8.0–10.0	–	0.5 max	0.3	0.15		500	
			2D	2.8–3.2	1.5–2.2	0.2–0.8	4.0–6.0	8.0–10.0	–	0.5 max	0.3	0.15		550	
			2E	3.2–3.6	1.5–2.2	0.2–0.8	4.0–6.0	8.0–10.0	–	0.5 max	0.3	0.15		600	

Notes: This table is for comparison purposes only. The National Standards should be referred to for details.

[1] Ni may be partly replaced by Cu.
[2] Supply as-cast only when no requirements are made about impact strength.
[3] In castings with section thickness greater than 125 mm, HB may be 50 less.
[4] Optional or, if necessary, as a combination.

Table 8.8 European specifications for high Cr-Mo abrasion resistant alloys

Country	Specification	Grade	C	Si	Mn	Ni	Cr	Cu	Mo	P	S
France	NF A 32–401 1980	FB Cr15MoNi	2.0–3.6	0.2–0.8	0.5–1.0	0.0–2.5	14.0–17.0	–	0.5–3.0	–	–
Germany	DIN 1695 1981	G-X300CrMo15 3	2.3–3.6	0.2–0.8	0.5–1.0	0.7[1]	14.0–17.0	–	1.0–3.0	–	–
		CrMoNi15 2 1	2.3–3.6	0.2–0.8	0.5–1.0	0.8–1.2[1]	14.0–17.0	–	1.8–2.2	–	–
UK	BS 4844 1986	3A	2.4–3.0	1.0 max	0.5–1.5	0.0–1.0	14.0–17.0	0–1.2	0.0–2.5	0.1	–
		3B	3.0–3.6	1.0 max	0.5–1.5	0.0–1.0	14.0–17.0	0–1.2	1.0–3.0	0.1	0.1
France		FB Cr20MoNi	2.0–3.6	0.2–0.8	0.5–1.0	0.0–2.0	11.0–14.0	–	0.5–3.0	–	–
Germany		G-X260CrMoNi 20 2 1	2.3–2.9	0.2–0.8	0.5–1.0	0.8–1.2[1]	17.0–22.0	–	1.8–2.2	–	–
UK		3C	2.2–3.0	1.0 max	0.5–1.5	0.0–1.5	17.0–22.0	2.0 max	0.0–3.0	0.1	0.06
France		FB Cr26MoNi	1.5–3.5	0.2–1.2	0.5–1.5[2]	0.0–2.5	22.0–28.0	0.0–1.5	0.5–3.0	–	–
Germany		G-X260Cr27	2.3–2.9	0.5–1.5	0.5–1.5	1.2	24.0–28.0	–	1.0	–	–
		G-X300CrMo27 1	3.0–3.5	0.2–1.0	0.5–1.0	1.2	23.0–28.0	–	1.0–2.0	–	–
UK		3D	2.4–2.8	1.0 max	0.5–1.5	0–1.0	22.0–28.0	2.0 max	0–1.5	0.1	0.06
		3E	2.8–3.2	1.0 max	0.5–1.5	0–1.0	22.0–28.0	2.0 max	0–1.5	0.1	0.06

Notes: This table is for comparison purposes only. The National Standards should be referred to for details.
[1] can also be produced as a Cu-containing alloy.
[2] in order to obtain a hard austenitic structure, Mn may be 4.0% max.

Table 8.9 USA Specifications for abrasion resistant cast irons: ASTM A532/A532M-93a

Class	Type	Designation	C	Mn	Si	Ni	Cr	Mo	Cu	P	S
1	A	Ni-Cr-Hc	2.8–3.6	2.0	0.8	3.3–5.0	1.4–4.0	1.0	–	0.3	0.15
1	B	Ni-Cr-Lc	2.4–3.0	2.0	0.8	3.3–5.0	1.4–4.0	1.0	–	0.3	0.15
1	C	Ni-Cr-GB	2.5–3.7	2.0	0.8	4.0	1.0–2.5	1.0	–	0.3	0.15
1	D	Ni-HiCr	2.5–3.6	2.0	2.0	4.5–7.0	7.0–11.0	1.5	–	0.1	0.15
11	A	12%Cr	2.0–3.3	2.0	1.5	2.5	11.0–14.0	3.0	1.2	0.1	0.06
11	B	15%Cr-Mo	2.0–3.3	2.0	1.5	2.5	14.0–18.0	3.0	1.2	0.1	0.06
11	D	20%Cr-Mo	2.0–3.3	2.0	1.0–2.2	2.5	18.0–23.0	3.0	1.2	0.1	0.06
111	A	25%Cr	2.0–3.3	2.0	1.5	2.5	23.0–30.0	3.0	1.2	0.1	0.06

The Cr–Fe oxides formed during melting attack silica lining materials, so alumina linings should be used. Charges are based on steel scrap and foundry returns with additions of high and low carbon ferrochromium to obtain iron of the required carbon content. Ferromolybdenum is added as needed. A loss of about 5% of the Cr content will occur during melting.

A melting temperature of 1600–1650°C is needed for the low carbon, high Cr irons, but a lower temperature of 1550–1600°C is sufficient for the higher carbon alloys.

A viscous oxide film forms on the surface of the molten metal in the furnace and the ladle. The iron must be carefully skimmed before pouring, and running systems must be designed to trap oxide dross. Metal filters should be used if possible. To avoid oxide defects on the casting surface, gating should provide rapid filling of the mould with minimum turbulence.

Chapter 9
Types of steel castings

Introduction

Steel is the most versatile of the structural engineering materials, so that more grades of steel are available for casting than any other alloy type. The casting of steel presents a number of special problems due mainly to the high casting temperature, typically 1550–1600°C, and the volume shrinkage (6.0–10.0%) that occurs on freezing. Steel castings range in size from a few grams to hundreds of tonnes. Steel castings are often used in critical situations where optimum mechanical properties are essential so that freedom from casting defects is particularly important. The advantages of cast steel over other cast metals, such as ductile iron, are:

High modulus – minimising deflection under heavy static loading
Shock and impact resistance – to cope with severe dynamic loading
Heat-treatable – to give the required toughness, hardness and strength
Weldable – so that castings can be joined to wrought steel products
High temperature strength – special alloys are available

Steel castings compete in many areas with steel weldments. In welding, it is difficult to avoid stress concentrations at junctions while castings have well-radiused joints so that fatigue resistance of cast steel junctions is superior to that of welded junctions. Also welds, especially in large structures, are not always amenable to stress-relieving after welding so that internal stresses are likely to be greater in weldments than in steel castings. This may lead to distortion of weldments or even failure in highly stressed components. Castings allow freedom of design which is not possible with weldments.

The main applications of steel castings, roughly in order of tonnage used in the UK, are:

Valves, pumps and compressors
Construction and earth moving equipment
Crushing, grinding, quarrying and dredging
Oil and gas rigs
Mining equipment
Motor vehicles
Defence equipment
Railways

Shipbuilding
Iron and steelworks
Nuclear power plant
General engineering equipment
Electrical plant
Agricultural equipment

While the overall tonnage of steel castings has been on a downward trend for a number of years, the overall value has increased due to a growing trend towards increased quality standards, reflecting the high standards needed for nuclear plant, gas and oil rigs etc.

Processes used for steel castings are:

Green sand	for high volume castings
Chemically bonded sand	where numbers required are smaller
Ceramic block moulds	for castings requiring particularly good surface finish
Lost wax investment casting	for smaller castings made in large numbers where good surface finish and dimensional accuracy is needed

The Lost Foam process is not normally used for steel casting because of the danger of carbon pick-up from the foam pattern.

Specifications for steel castings

Steel casting specifications in Europe are undergoing major changes as new European specifications are being drawn up. At the time of writing, the only European standard to be issued in the UK is BS EN 10213:1996 covering Steel Castings for Pressure Purposes. During the next few years, EN standards covering steels for General Engineering, Corrosion Resisting, Heat Resisting, Wear Resisting, Structural and Centrifugal applications will replace the existing national standards in European countries.

Until the new standards are issued, individual national standards apply. However, attention should be paid to the International Standard ISO 4990-1986 Steel Castings, General Technical Delivery Requirements which specifies the methods of testing steel castings and other matters which need to be agreed between purchaser and supplier. The contents of this standard are being incorporated into the forthcoming European standard.

ISO 4990 requires the purchaser to supply details with the enquiry covering:

A description of the casting(s) by pattern number or drawing
Dimensional tolerances, machining allowances and datum points for machining
The material standard and grade of steel

The type of document required for inspection and testing
Additional information such as agreed manufacturing method
Size of test lot
Procedures for marking, machining, protection, packaging, loading, dispatching and the destination
Submission of sample castings
Methods of statistical control
Inspection procedures

The standard lists details of inspection and testing to be carried out by the manufacturer including:

Documents required
Sampling, preparation of test pieces and mechanical and chemical test methods
Test blocks
Heat treatment
Repair welding
Non-destructive tests
Special tests such as intergranular corrosion and pressure tests
Test reports must provide the required traceability of the castings they represent

It is normally required that the manufacturer shall maintain records of results of all the chemical and mechanical tests performed by the foundry for a minimum of 5 years.

Other ISO Standards detail testing methods:

ISO 148	Steel – Charpy impact test (V-notch).
ISO/R 783	Mechanical testing of steel at elevated temperatures – Determination of lower yield stress and proof stress and proving test.
ISO 2605/1	Steel products for pressure purposes – Derivation and verification of elevated temperature properties – Part 1: Yield or proof stress of carbon and low alloy steel products.
ISO 2605/2	Steel products for pressure purposes – Derivation and verification of elevated temperature properties – Part 2: Proof stress of austenitic steel products.
ISO 3452	Non-destructive testing – Penetrant inspection – General principles.
ISO 3651/2	Austenitic stainless steels – Determination of resistance to intergranular corrosion – Part 2: Corrosion test in a sulphuric acid/copper sulphate medium in the presence of copper turnings (Monypenny Strauss test).
ISO 5579	Nondestructive testing – Radiographic examination of metallic materials by X- and gamma rays – Basic rules.
ISO 6506	Metallic materials – Hardness test – Brinell test

ISO 6892   Metallic materials – Tensile testing.
ISO 8062   Castings – System of dimensional tolerances.

British Standards for steel castings

A number of British Standard Specifications cover steel castings. There are equivalent standards in all industrialised countries, but the number of specifications is so great that it is not possible to list them here. The British Standards for steel castings include:

BS 1504   Steel Castings for Pressure Purposes
BS 1965 Pt 2   Butt Welding Pipe Fittings for Pressure Purposes (Austenitic Stainless Steel)
BS 3100   Steel Castings for General Engineering Purposes
                 Section 2: Carbon and Carbon–Manganese Steels
                 Section 3: Low Alloy Steel Castings
                 Section 4: Corrosion Resisting, Heat Resisting and High Alloy Steel Castings
BS 3146 Part 1   Investment Castings in Metal
                 Carbon and Low Alloy Steels
                 Corrosion and Heat Resisting Steels, Nickel and Cobalt Base Alloys
                 Vacuum Melted Alloys
BS 2772   Iron and Steel for Colliery Haulage and Winding Equipment
                 1.5% Mn Steel Castings for Mine Car Couplers
BS 4473   Railway Rolling Stock Material
                 Cast Wheel Centres in Unalloyed Steel for Tyred Wheels
BS 4534   Weldable Cr-Ni Centrifugally Cast Steel Tubes
BS Aerospace Series Specifications

Lists of equivalent standards are given in *British and Foreign Specifications for Steel Castings*, published by the Castings Development Centre (formerly the Steel Castings Research and Trade Association) East Bank Road, Sheffield S2 3PT, 4th Ed. 1980. The large number of specifications may be judged from the fact that the above book lists over 180 British specifications alone.

Types of steel castings

It is convenient to classify steels into three groups

    Carbon steels
    Low and medium alloy steels
    High alloy steels

Carbon steels

These are steels in which carbon is the main alloying element, other elements such as manganese and silicon are present for the purposes of deoxidation. The effectiveness of heat treatment in increasing the strength of the steel depends on carbon content and carbon steels may be further subdivided into

Low carbon steel	<0.20%C
Medium carbon steel	0.20–0.50%C
High carbon steel	>0.50%C

Applications of carbon steel castings include

Ship's structural castings
Railway rolling stock
Automotive castings
Hot metal ladles
Rolling mill equipment
Rolls and rollers
Machines and tools
Mine and quarry equipment
Oil and petroleum equipment

Typical medium carbon steel specifications are shown in Table 9.1 (from BS 3100). There are equivalent specifications in other countries, for example:

UK BS 3100	A1	A2	A3
USA ASTM	A27-77 N-1	A27-77 N-2	A148-73 gr. 80-40
Germany DIN 1681	GS-45	GS-52	GS-60

Carbon steels are most frequently used in the annealed, normalised or annealed and normalised conditions, though other heat treatments may be used to develop special properties, for example, for magnetic permeability.

Low alloy and medium alloy steels

Castings in this class normally contain amounts of manganese or silicon greater than is needed for deoxidation (e.g. >1% Mn) or to which other elements have been specially added. Alloy steels will contain Ni, Cr or Mo as main alloying constituents and smaller amounts of V, Cu and possibly B. These elements are added primarily to alter hardenability so that the required properties can be obtained in different section thicknesses.

Pearlitic manganese steels (1.5%Mn), Table 9.1, are used where strengths a little higher than attainable in plain carbon steels together with good

Table 9.1 BS 3100:1976 Steel Castings for General Engineering Purposes Carbon and Carbon–Manganese Steel Castings

| | Carbon Steel Castings for General Purposes | | | | | | $1\frac{1}{2}$% Manganese Steel Castings for General Purposes | | | | | |
| | A1* | | A2* | | A3* | | A4 | | A5 | | A6 | |
Chemical composition (%)	min	max	min	max	min	max	min	max	min	max	min	max
C	–	0.25	–	0.35	–	0.45	0.18	0.25	0.25	0.33	0.25	0.33
Si	–	0.60	–	0.60	–	0.60	–	0.60	–	0.60	–	0.60
Mn	–	0.90(1)	–	1.0 (1)	–	1.0 (1)	1.2	1.6	1.2	1.6	1.2	1.6
P	–	0.06	–	0.06	–	0.06	–	0.05	–	0.05	–	0.05
S	–	0.06	–	0.06	–	0.06	–	0.05	–	0.05	–	0.05
Cr	–	0.25 (2)	–	–	–	–	–	–	–	–	–	–
Mo	–	0.15 (2)	–	–	–	–	–	–	–	–	–	–
Ni	–	0.40 (2)	–	–	–	–	–	–	–	–	–	–
Cu	–	0.30 (2)	–	–	–	–	–	–	–	–	–	–
Mechanical properties												
TS (N/mm²)	430	–	490	–	540	–	540	690	620(5)	770(5)	690(6)	850(6)
0.2%PS (N/mm²)	230	–	260	–	295	–	320	–	370(5)	–	495(6)	–
Elongation (%)	22	–	18	–	14	–	16	–	13(5)	–	13(6)	–
Angle of bend (°)	120 (3)	–	90 (3)	–	–	–	–	–	–	–	–	–
Radius of bend (t)#	1.5 (3)	–	1.5 (3)	–	–	–	–	–	–	–	–	–
Charpy impact (J)	25 (3)	–	20 (3)	–	18 (3)	–	30	–	25(5)	–	25(6)	–
Hardness (HB)	–	–	–	–	–	–	152	207	179(5)	229(5)	201(6)	255(6)

Notes: *If attached test samples are used, mechanical properties expected only where max. section thickness of casting is < 500 mm
t = thickness of the test piece.
(1) For each 0.01%C below the max., an increase of 0.04%Mn will be permitted up to a max. of 1.10.
(2) Residual elements: the total shall not exceed 0.80%.
(3) Either a bend test or an impact test may be specified.
(4) Impact test mandatory only if specified by purchaser.
(5) Limiting section thickness, 100 mm.
(6) Limiting section thickness, 53 mm.
This table is intended only as a guide, refer to the British Standard for details.

toughness are required. Applications are excavating buckets, travelling wheels, gears, tractor track links and pneumatic tool parts.

Frequently used alloy steels are $1^1/_2$% Mn-Mo, $1^1/_4$% Cr-Mo, 0.75% Ni-Cr-Mo, $1^1/_2$% Ni-Cr-Mo and 3% Cr-Mo (Table 9.2). Such higher strength steels find use for gears, shaft pinions, impellers, rolls, rollers, crusher heads, mill liners, crankshafts, cement processing machinery, fittings and components in the oil industry.

Steel castings with a hard skin and a very tough core may be obtained by local hardening (flame or induction hardening or carburising or nitriding).

Austenitic manganese steel

This widely used steel contains from 10–14%Mn and 1.0–1.4%C (typically 12%Mn, 1.25%C) (Table 9.3). The as-cast structure contains carbides and other transformation products which produce brittleness. The castings are therefore austenitised at 1000°C or above and quenched in water, producing a fully austenitic structure that is quite tough. On account of their unique ability to resist pounding and severe abrasive wear, Mn-steel castings are used for railway trackwork, dredging, excavating, pulverising and crushing equipment for mines and quarries, cement mills etc.

High alloy steels

These are used where corrosion resistance or heat resistance is required. The alloys may be subdivided into three main groups:

Fe-Cr
Fe-Cr-Ni
Fe-Ni-Cr

The iron-chromium steels contain little or no nickel and 10–30%Cr. The Cr content influences their resistance to scaling, the 13%Cr steels having reasonable oxidation resistance up to 650°C while 28%Cr steel can be used up to around 1100°C. The 13%Cr steels are used for pumps and steam turbine equipment. 20–30%Cr steels are used in the chemical industry for pump parts and valves, also for furnace parts, grates, kiln parts etc.

The iron-chromium-nickel alloys include the well-known 18/8 stainless steels to which additions of Nb or Ti may be used to stabilise the austenite on heating. Mo may also be added and Ni raised to 10% to allow the steel to be used at higher temperatures (up to 1100°C) and improve the corrosion resistance. High corrosion resistance and toughness require the lowest carbon possible and special melting techniques allow figures as low as 0.03%C to be achieved, although achieving such low carbon contents at the surface of the casting is not easy since most moulding materials can cause surface

Table 9.2 BS 3100:1976 Steel Castings for General Engineering Purposes Low Alloy Steel Castings

Chemical composition (%)	$C-\frac{1}{2}Mo$ B1		$1\frac{1}{4}Cr-Mo$ B2		$2\frac{1}{4}Cr-Mo$ B3		3%Cr–Mo B4		$1\frac{1}{4}\%Cr-Mo$ BW4	
	min	max	min	max	min	max	min	max	min	max
C	–	0.20	–	0.20	–	0.18	–	0.25	0.55	0.65
Si	0.20	0.60	–	0.60	–	0.60	–	0.75	–	0.75
Mn	0.50	1.00	0.50	0.80	0.40	0.70	0.30	0.70	0.50	1.00
P	–	0.05	–	0.05	–	0.05	–	0.04	–	0.06
S	–	0.05	–	0.05	–	0.05	–	0.04	–	0.06
Cr	–	0.25	1.00	1.50	2.00	2.75	2.50	3.50	0.80	1.50
Mo	0.45	0.65	0.45	0.65	0.90	1.20	0.35	0.60	0.20	0.40
Ni	–	0.40	–	0.40	–	0.40	–	0.40	–	–
Cu	–	0.30	–	0.30	–	0.30	–	0.30	–	–
Mechanical properties										
TS (N/mm²)	460	–	480	–	540	–	620	–	–	–
0.2%PS (N/mm²)	260	–	280	–	325	–	370	–	–	–
Elongation (%)	18	–	17	–	17	–	13	–	–	–
Angle of bend	120	–	120	–	120	–	120	–	–	–
Radius of bend (t)	1.5t	–	1.5t	–	3t	–	3t	–	–	–
Charpy Impact (J)	20	–	30	–	25	–	26	–	–	–
Hardness (HB)	–	–	140	212	156	235	179	255	341	–

Notes:
t = thickness of the test piece
This table is intended only as a guide, refer to the British Standard for details.

Table 9.3   BS 3100:1976 Steel Castings for General Engineering
Purposes
Austenitic Manganese Steel Castings

	BW10	
Chemical composition (%)	*min*	*max*
C	1.00	1.25
Si	–	1.00
Mn	11.00	–
P	–	0.07
S	–	0.06
Cr	–	–
Mo	0.45	–
Ni	–	–
Cu	–	–

The mechanical properties shall be agreed between the manufacturer
and the purchaser.
For special applications, max. carbon may be increased to 1.35%.
This table is intended only as a guide, refer to the British Standard
for details.

recarburisation due to the breakdown of carbon-containing binders. These
corrosion resistant steels are widely used for marine fittings, pump parts,
valve bodies, impellers etc.

One of the most widely used cast stainless steels is BS 3100: 316C16
(Table 9.4) an 18Cr/10Ni/2.5Mo cast equivalent of 316 wrought stainless
alloy (US equivalent, CF-8M).

Table 9.4   Cast stainless steel BS3100: 316C16

C	Si	Mn	P	S	Cr	Mo	Ni
0.08	1.5	2.0	0.04	0.04	17–21	2.0–3.0	10 min

Mechanical properties	
UTS (N/mm^2)	480
0.1%PS (N/mm^2)	240
Elongation (%)	26
Impact (J)	34

(Refer to BS3100 for details)

Such steels can be used at temperatures up to about 550°C making them
suitable for high pressure steam valves for nuclear and chemical applications
as well as for valves and pumps for handling hot acids and also marine
fittings. Unlike the wrought alloys, cast stainless steels contain from 5–40%
ferrite which provides strength, improves weldability and maximises
resistance to corrosion in specific environments.

Alloys for heat resisting applications contain 25%Cr and 12%Ni for use
up to 1050°C and 25%Cr–20%Ni for use up to 1100°C. They are used for
exhaust manifolds, burner nozzles, furnace parts, stack flues etc.

The iron-nickel-chromium alloys are used for heat resisting applications. Widely used alloys include one containing 17%Cr, 25%Ni, and one having 37%Ni, 15%Cr. The castings are used for furnace parts, salt and cyanide pots, annealing trays etc.

Typical steel casting compositions of the above steel types are given in Table 9.5.

Duplex steels

The severe conditions experienced in off-shore drilling platforms and other marine applications require high strength corrosion resistant steels. Alloys having a duplex ferrite-austenite microstructure have been developed for this purpose (Table 9.6). High hardness is produced by the high copper content (3%). The steel is heat treated by austenitising at around 1100°C when the copper dissolves. Oil quenching retains the copper in solution, subsequent ageing at around 480°C produces a ferrite-austenite structure with precipitated copper increasing strength and hardness.

Only a few specifications are quoted here. Details of Steel Casting Specifications world-wide are presented in *British and Foreign Specifications for Steel Castings*, published by CDC, East Bank Road, Sheffield S2 3PT. The large number of specifications may be judged from the fact that the above book lists over 180 British specifications alone.

Further details of the properties of all the above steels may be obtained from *Steel Castings Design Properties and Applications*, edited by W.J. Jackson, published by CDC, East Bank Road, Sheffield S2 3PT.

The Steel Castings Handbook, sixth edition, by Steel Founders' Society of America and ASM International, 1995 contains full details of material selection, mechanical properties, design and manufacture of castings etc.

Physical properties of steels

Table 9.7 gives typical physical properties of commonly used cast steels. These figures should be regarded as approximate only, since the exact composition and state of heat treatment affects the properties. Where properties are quoted for a particular temperature, the figure is the average value from room temperature to the quoted temperature.

Selection of suitable steel for casting

The choice and specification of a suitable steel for a specific application is far from easy. Not only is it necessary to specify an alloy having suitable properties: strength, hardness, fatigue properties, low temperature strength,

Table 9.5 BS 3100:1976 Steel Castings for General Engineering Purposes Alloy Steel Castings

Chemical composition (%)	13%Cr 420C29 min	max	25–30%Cr 452C12 min	max	18/10CrNi 316C16 min	max	25/20CrNi 310C40 min	max
C	–	0.20	1.00	2.00	–	0.08	0.30	0.50
Si	–	1.00	–	2.00	–	1.50	–	1.50
Mn	–	1.00	–	1.00	–	2.00	–	2.00
P	–	0.04	–	0.06	–	0.04	–	0.04
S	–	0.04	–	0.06	–	0.04	–	0.04
Cr	11.50	13.50	25.00	30.00	17.00	21.00	24.00	27.00
Mo	–	–	–	1.50	2.00	3.00	–	1.50
Ni	–	1.00	–	4.00	10.00	–	19.00	22.00
Cu	–	–	–	–	–	–	–	–
Mechanical Properties								
TS (N/mm²)	690	–	480	–	480	–	620	–
0.2%PS (N/mm²)	465	–	280	–	240	–	370	–
Elongation (%)	11	–	17	–	26	–	13	–
Angle of bend	–	–	120	–	–	–	120	–
Radius of bend (t)	–	–	1.5t	–	–	–	3t	–
Charpy Impact (J)	–	–	30	–	34	–	26	–
Hardness (HB)	201	255	140	212	–	–	179	255

This table is intended only as a guide, refer to the National Standards for details.

Table 9.6   Duplex steel GX2CrNiMoCN25-6-3-3

C	Si	Mn	P	S	Cr	Mo	Ni	Cu	N
0.03	1.0	1.50	0.035	0.025	24.5–26.5	2.5–3.5	5.0–7.0	2.75–3.5	0.12–0.22

Mechanical properties:
UTS (N/mm^2)   650–850
0.1%PS (N/mm^2)   480
Elongation (%)   22
Impact (J)   50

Table 9.7   Physical properties of some steels

Composition	Temp. °C	Density (g/cm^3)	Specific heat capacity (J/kg.K)	Thermal expansion coeff. (10^{-6}/K)	Thermal conductivity (W/m.K)	Electrical resistivity (micro-ohms.m^2/m)
Carbon steels	RT	7.86			59.5	13.2
	100		482	12.19	57.8	19.0
	200		523	12.99	53.2	26.3
	400		595	13.91	45.6	45.8
	600		741	14.68	36.8	73.4
	800		960	14.79	28.5	108.1
	1000			13.49	27.6	116.5
Low alloy steels	RT	7.85				25.4
	100		456	12.45	48.2	30.6
	200		477	13.2	45.6	39.1
	400		532	14.15	39.4	60.0
	600		599	14.8	33.9	88.5
Stainless 18/8	RT	7.92			15.9	69.4
	100		511	14.82	16.3	77.6
	200		532	16.47	17.2	85.0
	400		569	17.61	20.1	97.6
	600		649	18.43	23.9	107.2
	800		641	19.03	26.8	114.1
	1000				28.1	119.6
13%Cr	100	7.75	482	11.0	24.3	57.0

wear resistance, corrosion resistance, high temperature strength and resistance to oxidation etc. but it is usually necessary to define the inspection procedures to be used. These will include

Visual examination of the casting surface to determine surface roughness and freedom from cracks
Magnetic particle surface examination
Liquid penetrant surface examination
Radiographic inspection, usually in comparison to a system of reference radiographs

Ultrasonic testing, where heavy sections prohibit the use of radiography
Pressure or leak testing
Dimensional inspection
Statistical process control
Quality Assurance usually in accordance with a standard such as ISO 9002 or ISO 4990 Steel Castings – General Technical Delivery Requirements

The cost of minimum specification requirements is included in the basic casting price, but additional testing will involve additional costs which must be agreed between customer and supplier before castings are ordered.

Chapter 10

Melting and treatment of steel for casting

Large steel foundries may use electric arc furnaces but induction furnaces are the most commonly used melting furnaces for making steel castings. Arc furnaces are capable of using low cost scrap charges, since refining takes place in the furnace but arc furnaces have limitations since there is always some pick-up of carbon from the graphite electrodes, so that very low carbon stainless steels (<0.03%C) cannot be made. Refining is not possible in the induction furnace, so a carefully selected charge must be used, but any type of steel may be melted.

Arc furnace melting

The furnace consists of a refractory-lined mild steel shell, circular in section and having a dished bottom. The refractory roof, usually water-cooled, has three graphite electrodes projecting through it into the charge. The electrodes can be raised and the roof lifted hydraulically and swung aside to allow the furnace to be charged by a drop bottom charge bucket. The furnace is tilted forward for tapping and backward for slag removal (Fig. 10.1).

Electrode arms above the roof supply three-phase power to the electrodes and allow each electrode to be independently raised or lowered mechanically. The length of the arc governs the power input to the furnace, regulators measure the current in each electrode and control it automatically by continuously adjusting the position of the electrode to maintain an optimum arc length.

500–600 kWh of electricity is required to melt and raise to casting temperature 1 tonne of steel. Furnaces are usually rated at 500 kVA per tonne to give a melting time of about $1^1/_2$ hours.

The electrodes are made of graphite and are consumed during operation through oxidation, volatilisation and breakage so they must be replaced as necessary. A 3 tonne furnace typically uses electrodes of 200 mm diameter. Electrode consumption is a significant factor in the cost of arc melting and figures vary from 3–10 kg/tonne of steel melted depending on the type of steel being produced and the practice used.

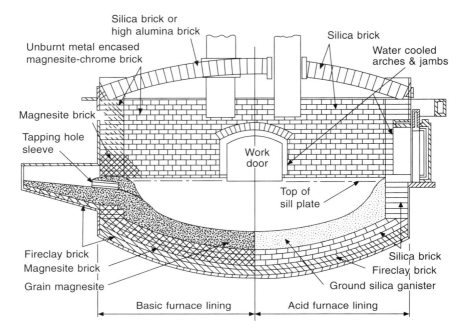

Figure 10.1   *Schematic diagram of an arc furnace showing typical refractories used in acid and basic practices. (From Jackson, W.J. and Hubbard, M.W., Steelmaking for Steelfounders, 1979, SCRATA. Courtesy CDC.)*

Refractories

Arc furnace refractories can be 'acid', 'basic' or 'neutral' depending on the melting practice used. Basic linings of dolomite or magnesite are the most commonly used in steel foundries. A safety lining of firebrick is installed in the steel shell of the furnace, followed by an inner lining of magnesite or dolomite bricks. The working hearth is then formed by ramming tarred dolomite or magnesite. Side walls are built up from tarred dolomite or steel-clad magnesite bricks. The lining is then dried and fired. Roofs are expected to have a long, maintenance-free life and may be made of high alumina refractory.

Melting practice in the basic lined arc furnace

The use of basic slags, oxidising or reducing in nature, allows phosphorus and sulphur to be reduced to low levels so that the basic lined furnace is very versatile, allowing good quality steel to be made from poor quality scrap.

Melting in a basic arc furnace can be a one or two stage process using first an oxidising slag then, if necessary, a reducing one. The charge, usually of foundry returns and purchased steel scrap, is made up in a drop-bottom charging bucket adding limestone and sometimes iron ore to the charge to make the first oxidising slag which will remove phosphorus. Prior to the scrap being dropped into the furnace, some form of recarburiser such as anthracite, coke or crushed electrodes is placed on the furnace bottom to ensure sufficiently high carbon content at melt-out to give an adequate 'boil'. When the charge is melted, the boil is induced by adding iron ore or gaseous oxygen to the melt. The oxygen reacts with carbon in the melt forming bubbles of carbon monoxide which rise through the metal giving the appearance of boiling. The vigorous stirring action of the boil ensures effective reaction of the metal with the slag and removes hydrogen and nitrogen from the melt. The boil also removes carbon from the melt. The carbon content of the melt is taken below the required finishing carbon and the boil stopped by adding a deoxidant such as silicon and/or aluminium. The slag is then removed, taking the phosphorus from the charge with it.

If sulphur removal is required, a second, reducing slag is formed by adding limestone with fluorspar to keep it fluid and reducing agents such as powdered ferrosilicon and calcium carbide. The chemistry of the melt is adjusted by adding carburiser, Fe-Si, Fe-Mn together with Fe-Cr and Fe-Mo if required. During time under the reducing slag hydrogen and nitrogen pick-up can occur, so the reducing period should be as short as possible. The slag is removed before pouring the steel into the ladle. The steel is finally deoxidised in the ladle with aluminium.

There are many variations of the basic arc steelmaking process. The practice used depends on the type of steel to be made and the quality of the scrap steel used for the charge.

Induction furnace melting

The medium frequency coreless induction furnace (Fig. 4.5) is the most flexible steel melting unit and is widely used in steel foundries. Small sized scrap can be melted from cold and the furnace can be used for successive melts of widely differing analyses. Melting is rapid since high power can be applied. There is a strong stirring action in the molten steel due to the induced current, so that a slag cover cannot be maintained. For this reason there is no refining stage, the liquid steel having similar composition to the charged scrap except for carbon, silicon and manganese which are lost by oxidation and must be made up to the required levels by adding pig iron, or other carburisers and ferroalloys. Alloying elements may also be added and adjustments to temperature can be rapidly made. High alloy steels are frequently made in the induction furnace since the melting losses of expensive alloying elements are low. Extra-low carbon stainless steels can be melted without pick-up of carbon.

Design of the furnace

The principles of design and operation of coreless induction furnaces are described in Chapter 4, pp. 55–60.

Induction furnaces for steel melting may have a melt capacity of up to 6 tonnes. Frequencies of 1000 Hz are usually used up to about 3 tonnes and 500 Hz for larger furnaces, being powered at up to 750 kW per tonne of capacity. Such furnaces can be started from cold and have a charge-to-tap time of about 1 hour.

Refractories

The linings for induction furnaces are nowadays usually made of magnesium-alumina spinel, which can expand or contract on firing. The usual grade is 30% alumina, 70% magnesia which has an expansion sufficient to produce metal tightness without damaging the coil.

Linings are made from refractory ramming mixes. The grading of the refractory is important since it controls packing density and sintering. A grading of around 45% coarse, 10% medium and 45% fines gives the maximum packing density. The fines control the sintering of the bond which may be formed by additions of boric acid or boric oxide, often incorporated in a proprietary ramming mix. Before installing the lining, the coil and its insulation should be checked for damage. The usual practice with medium frequency furnaces is to coat the copper coil of the furnace with a layer of 'mudding' about 6 mm thick, of a medium to high alumina cement. This remains in place when the hot face is knocked out. Between this and the hot face is a layer of ceramic fibre insulation. The furnace bottom is first formed by pouring in refractory to a depth of 50–60 mm which is rammed to give the correct depth. The working face is formed by compacting the refractory behind a steel former concentrically placed within the coil. Formers are normally constructed from mild steel sheet according to the furnace manufacturer's design. Refractory is poured between the former and the coil and compacted using vibratory ramming tools or manual compaction adding more refractory as compaction proceeds. The top layer of the lining is usually finished by mixing the granular lining refractory with sodium silicate.

The former is left in the furnace and melted out with the first heat. Power is slowly increased giving time for the bond to form before the former melts.

Charge materials

The composition of steel melted in the induction furnace changes little during melting so that careful selection of charge materials is necessary. Scrap should be clean, often shot blasted to remove sand, and free from oil and moisture which may cause hydrogen pick-up and fume. The scrap is charged together with non-oxidisable alloys. When melted, the oxidisable alloys are added and the melt brought to the required temperature then tapped as quickly as possible. Holding the melt at high temperature should be avoided to minimise hydrogen pick-up which occurs quite rapidly above 1600°C.

The recycling of scrap in an induction furnace can lead to the pick-up of gases since no refining takes place. Where possible, scrap from induction furnaces should be returned to arc furnaces for refining.

AOD refining

AOD refining (argon oxygen decarburisation) has been developed to process alloy steels for demanding applications. The metal is melted in a primary arc or induction furnace then transferred to a special AOD vessel with a controlled slag cover. Oxygen diluted with argon is then injected through a tuyere at the bottom of the vessel. By reducing the partial pressure of oxygen, carbon can be removed to low levels without oxidising the chromium and other elements in the melt. Removal of S and P by slag/metal reaction is aided by the vigorous stirring of the bath induced by the gas injection.

Steel with precise chemistry can be produced using AOD but at high cost since the extra vessel is expensive and refractories must be replaced frequently. It is not widely used in steel foundries outside the USA.

Melting and casting quality

Aspects of casting quality affected by melting and steelmaking practice include

Deoxidation effects
Sulphides
Inclusions
Gas porosity
Hot tearing

Deoxidation

Liquid steel dissolves oxygen; as the steel cools, the dissolved oxygen combines with carbon in the steel forming carbon monoxide:

$$[C] + [O] = CO \ (g)$$

If allowed to occur during solidification this would cause gas porosity in the castings. The amount of dissolved oxygen in the liquid steel depends on the carbon content of the melt, low carbon steel containing higher oxygen. The dissolved oxygen is also affected by the silicon and manganese content of the steel. To avoid gas evolution during solidification, the melt must be deoxidised with an element which combines preferentially with the dissolved oxygen forming a stable oxide. Silicon, calcium silicide, titanium, zirconium and aluminium are commonly used deoxidants with aluminium being the most powerful. Ferrosilicon, calcium silicide or aluminium may be used as a 'block' in the furnace before tapping but the use of aluminium for the final deoxidation of liquid steel is practically universal.

The addition is usually made in the ladle. Aluminium is commonly used in stick form for rapid solution. Additions vary from about 0.1% for medium carbon steel to 0.2% for low carbon steel. Recovery is between 35 and 80%. Mechanical feeding of aluminium wire is also used, precise control of the addition is possible since the wire is fed rapidly deep into the melt giving predictable yields. It is important to control the residual aluminium left in the melt after deoxidation. Too little may allow the formation of harmful forms of sulphide and intergranular AlN. Too much will harmfully affect the steel properties. While aluminium is a powerful deoxidant, the products of deoxidation are solid crystalline inclusions which are slow to separate from the molten steel. They reduce the fluidity of the steel and have a tendency to precipitate in the nozzle bore, reducing the rate at which steel discharges from the ladle.

CaO fluxes alumina and controls the shape and size of the oxides allowing them to float out more effectively in the casting ladle. It is difficult to introduce sufficient calcium into the steel because when a lump of calcium alloy contacts molten steel, most of the calcium boils off before it has a chance to enter the steel where it can be effective. Calcium silicide cored wire may be added to the ladle using a mechanical wire injector. In this way the calcium is introduced deep into the steel where the ferrostatic head suppresses the rapid vaporisation of calcium so that it is more effective than bulk additions. Removal of inclusions is more effective if the steel is 'rinsed' with argon after the calcium treatment through a porous diffuser in the ladle.

Sulphides

As the tensile strength of the steel increases, so the harmful effect of sulphur increases. In carbon steel castings, 0.06%S may be tolerated but in aircraft specifications sulphur is limited to 0.02%max. Careful selection of charge materials is necessary to ensure low sulphur in induction melted steels. Sulphur is soluble in liquid steel but on solidification it precipitates as MnS. The sulphides formed may be globular, chain and film, or angular and are known respectively as Types I, II or III. Type II sulphides are the most harmful and must be avoided since they seriously affect the Charpy V-notch values of the steel.

The form of the sulphides is related to the residual aluminium content of the steel after deoxidation. The least harmful Type I globular sulphides are formed when residual Al is very low, but this is difficult to control in practice. The harmful Type II sulphides form at intermediate residual Al levels while the less harmful Type III angular sulphides are formed at higher residual Al levels. The carbon content of the steel also affects the amount of residual Al needed to ensure Type III sulphides. Low carbon steel requires higher residual Al (Table 10.1).

Table 10.1   Residual Al% needed to ensure formation of Type III sulphides in steels of different carbon content

C (%)	Residual Al(%)
0.1	0.04
0.2	0.03
0.3	0.02
0.4	0.02
0.5	0.02

From Jackson & Hubbard p. 235.

Steels with carbon above 0.5% should be deoxidised with 0.04% Al to ensure 100% type III sulphides. The aluminium addition must be increased as the carbon decreases but the addition should not be too high or the impact properties may suffer. Steels with carbon contents below 0.2% may advantageously be deoxidised using Al plus calcium silicide.

Inclusions

Inclusions in steel castings can arise from slag entrapment or from the erosion of furnace or ladle linings and refractories. They also arise from deoxidation, being either direct deoxidation products or products arising from reaction of the killed steel with refractories. For example manganese in killed steel will reduce the silica in siliceous refractories, the MnO so formed has a severe fluxing action on firebrick refractories.

Reoxidation of steel during teeming is also a source of inclusions. It has been estimated that in each pouring or tapping operation, 1–4 kg of oxide are produced for each tonne of steel transferred. Systems of shrouding the metal stream with an inert gas have been developed but are not very effective. The use of calcium silicide in conjunction with controlled additions of aluminium and argon rinsing are considered effective in reducing inclusions but can be expensive. Filtration of liquid steel using ceramic foam filters, (see Chapter 18) is now widely used to reduce the presence of inclusions in steel castings.

Gas porosity

Gas porosity in steel castings can be attributed to one or both of the following reactions:

$$[C] + [O] = CO\ (g) \tag{1}$$

$$2[H] = H_2(g) \tag{2}$$

A reaction similar to (2) is possible for nitrogen.

To minimise porosity, the hydrogen and nitrogen contents of the liquid steel must be minimised and the carbon/oxygen reaction prevented by a sufficiently strong deoxidation practice.

Hydrogen

The hydrogen level at which gas holes occur varies according to deoxidation practice and steel composition. A casting can be made sound as follows:

> up to 9 ppm of hydrogen for silicon killed steel castings
> 9–13 ppm hydrogen for Si and Al killed steel castings
> over 13 ppm hydrogen, impossible to obtain solid castings

Hydrogen in liquid steel arises from contact with moisture, usually from damp refractories and additives. Excessive bath temperatures and prolonged refining periods should be avoided. All late additions should be dry and ladles should be thoroughly dried. High atmospheric humidity is a factor since it may lead to moisture pick-up from refractories (and later from the sand moulds). A new furnace lining may cause excessive hydrogen pick-up and it is good practice to melt a 'wash heat' before the first charge proper.

Nitrogen

The main source of nitrogen in steel is the atmosphere and the nitrogen content of the steel depends to a large extent on the length of time that the bath is in contact with the atmosphere. While nitrogen alone can contribute to porosity, its normal level is far below the limit of solid solubility. In 0.10%C plain carbon steel castings, when nitrogen is at the 0.013% level, there is a distinct tendency towards porosity. Nitrogen together with hydrogen can sometimes be the cause of porosity in high alloy steels, particularly those containing large amounts of chromium, if high nitrogen ferrochromium has been used.

Nitrogen pinholes occur in castings made in moulds made using nitrogen-containing resins such as UF-furanes and phenolic urethanes, and it is important to use low nitrogen binders, particularly when reclaimed sand is used (see Chapter 13).

In addition to the possibility of causing porosity, nitrogen in combination with aluminium may influence the ductility of the steel due to intergranular fracture caused by AlN precipitating at grain boundaries as the steel cools through the temperature range 800–900°C. In commercial steel castings, the minimum amounts of nitrogen combined as aluminium nitride that can give rise to intergranular fracture are 0.002% for low alloy steels and 0.004% for plain carbon steels. It is not generally practical to reduce nitrogen to such a low level. However, if there is a large enough excess of aluminium in the steel after deoxidation, AlN can form in the melt and act as nuclei for precipitation in the solid state. The large AlN particles so formed do not give rise to intergranular fracture.

Carbon monoxide

The more thorough the deoxidation of the steel, the less the pinholing tendency, and proper deoxidation with aluminium is the most effective control. There must be an excess of deoxidant to compensate for the higher oxygen content that develops in the surface layer of a casting as it comes in contact with the mould wall. It is possible for silicon and aluminium to be removed locally which may result in pinhole porosity.

Hot tearing

Hot tearing is a common foundry problem caused by the contraction of the steel on freezing being hindered by the mould or cores. Chemical composition of the steel affects hot tearing in a minor way, sulphur having the greatest effect. Lowering sulphur to below 0.02% can be beneficial in carbon and low alloy steels, and a manganese content of 0.7–1.3%Mn ensures that any sulphur present precipitates as MnS rather than the Fe-FeS eutectic which increases the tendency to hot tear. In high alloy steels, austenitic 18/8 steels do not tear as severely as plain carbon or 13%Cr steels. As in the case of carbon steels, sulphur has the greatest adverse effect and manganese improves tear resistance.

Since hot tearing is caused by mechanical restraint preventing metal shrinkage occurring, casting design changes, such as use of adequate fillets at junctions, can be of great help. The minimum binder should be used in cores and moulds for any casting susceptible to the problem.

The above description of the melting and treatment of steel for casting is far from complete and for further details the book *Steelmaking for Steelfounders,* by W.J. Jackson and M.W. Hubbard, published in 1979 by the Steel Castings Research and Trade Association, Sheffield is recommended. Thanks are due to the Castings Development Centre, 7 East Bank Road, Sheffield for permission to use data from this book and from *Steel Castings Design Properties and Applications* referred to above.

Chapter 11
Molten metal handling

Iron foundries

Metal handling systems

The aim of the metal handling system is to deliver clean liquid metal to the mould at the required temperature. Automatic pouring systems are used increasingly, but manual pouring is still the commonest method of filling moulds. Iron foundries almost always use lip-pouring ladles although bottom-pouring may be used for very large castings such as ingot moulds.

For efficient energy use, molten metal tapping temperature should be as near as practicable to the desired pouring temperature, having regard to unavoidable heat losses. In general this is achieved by:

Use of the largest ladle size practicable for pouring moulds.
Ladles should always be covered.
Use of transfer ladles should be avoided, if possible.
Avoid intermittent use of pouring ladles.
Use efficient ladle-heating to prepare ladles for use and when unavoidable delays occur.
Bucket type ladles with insulated covers lose less heat than drum-type ladles.

Each transfer of liquid iron from one ladle to another results in a temperature loss of about 10–20°C.

Ladle lining

The ladle lining in an iron foundry is normally expected to last for a full shift, in which time it may be used between 50 and 100 times. The lining material must therefore be of good quality and be applied carefully. Traditionally, iron foundry ladles have been lined with ganister (crushed silica rock), naturally bonded sand or firebrick or a combination of these materials.

Ganisters are usually pre-mixed with a clay binder and are applied with the minimum amount of water, using a removable former of metal or wood.

If ladles are used daily on high production plants, the old lining should be removed completely from the ladle body, which should then be coated lightly with a ganister wash and the new lining material rammed into position round the former. The material must be properly vented, by hanging rods between the ladle shell and the lining. These are removed before drying to allow easy passage of steam from the lining.

Good quality naturally bonded sand may be used as an alternative to ganister but the bottom of the ladle should preferably be either brick or ganister lined to withstand the erosion during filling. The lip and rim of the ladles are also best lined with ganister. Again, venting is necessary to assist drying.

Larger ladles are lined with silica firebrick faced with a layer of ganister.

CO_2/silicate sand linings may also be used, they are applied using a former and cured by gassing. Because the free moisture is low, less drying is needed before use.

The greater the capacity of the ladle, the thicker should be the lining. Typical thicknesses are:

Ladle capacity (kg)	*Thickness* (mm)
500	25–50
500–5000	50–75
over 5000	75–150

Linings can be damaged by incorrect drying. Properly designed ladle heaters are better than simple gas-torches. The initial drying must be slow, to avoid cracking, the gas flame can be increased in intensity as the lining heats up. When the ladle is completely dry, it is ready for pre-heating. The lining temperature should be raised to about 900–1000°C before filling with molten metal. Heat losses are greatest when the ladle is filled for the first time after pre-heating or if it is only used intermittently, since heat from the metal is used in heating the refractories to temperature. If ladles are used continuously, temperature losses stabilise to a predictable value as heat is lost by radiation from the metal surface and by conduction through the ladle.

The actual cooling rate for any ladle depends on factors such as:

Capacity; larger ladles cool more slowly
Shape; narrow, deep ladles lose less heat by radiation from the metal surface
How full the ladle is; the fuller, the less the heat loss
The use of covers; a properly fitted cover can halve the cooling rate.

Typical temperature losses from open bucket ladles, well-lined and full are:

Capacity	25 kg	100 kg	500 kg	30 tonnes
Temp. loss	30°C/min	10°C/min	6°C/min	<2°C/min

These figures are only a rough guide, for greater accuracy, actual measurements should be made on the ladles in use. The use of a cover will approximately halve temperature losses.

Heat losses can be reduced by covering the surface of the metal with an insulating layer (note that it is not desirable to leave slag or dross on the surface because of the possibility of metal/slag reaction). RADEX 460L refractory insulator has been developed for this purpose, it is insulating and inert, having low carbon content and is virtually aluminium free. The metal should be skimmed clean and a layer of RADEX 460L scattered over the surface to give a layer about 10 mm thick, or roughly 1.5 kg/tonne of iron, though this depends on the ladle shape.

KALTEK ladle lining system for iron foundries

The KALTEK system uses a thermally insulating material as a liner to reduce the time taken to prepare foundry ladles and to provide consistent pouring temperatures (Fig. 11.1). Pre-heating of ladles is eliminated and installation is quick and easy. One-piece moulded KALTEK liners are available for many sizes of iron ladles. The lining is designed to fit accurately into the ladle shell. A permanent safety lining may sometimes be used as well. A coarse grained sand (10–20 AFS) is recommended to back-fill the lining to ensure adequate venting as the organic binders in the liner system burn out. A sodium silicate bonded core sand, adequately vented, can be used as a sand cap.

When molten metal is first poured into a cold ladle lined with KALTEK, heat is absorbed by the lining, which initially causes the metal to cool. On successive fills, the lining quickly reaches steady state and the cooling becomes less. Cooling of the metal in the ladle is less than in conventionally lined ladles because of the insulating properties of the KALTEK material. When used for iron casting, the life of the lining depends on the metal grade, slag type and pouring temperature but is usually several hundred fills, allowing multi-shift use. After use, the one piece lining is easily removed by breaking the sand cap and overturning the ladle. Cooling is usually unnecessary and the ladle can be relined immediately. To complement the KALTEK system, a one-piece insulating lid made from PROCAL can be used to reduce temperature losses even further.

Two-piece KALTEK ladle liners are available for sizes larger than can be moulded as a single liner. For even larger ladles, linings can be formed from KALTEK boards, supplied in the form of segments and bottom boards (Fig. 11.2). The ladle must be furnished with a permanent base lining of alumina or firebrick to act as a safety lining. The KALTEK segments and base boards are fitted inside the permanent lining, joints being filled with KALSEAL refractory cement. The gap between the KALTEK and the permanent lining is filled with coarse, dry silica sand, 15 AFS is recommended, or BAKFIL. Multiple filling is possible as long as the ladle lining is not allowed to cool down between fillings.

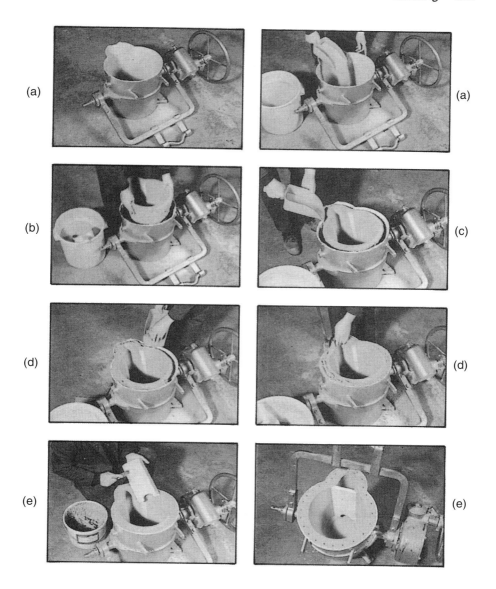

Figure 11.1 *Installation of the KALTEK one-piece ladle liner*
(a) Coarse grade sand (or BAKFIL) is placed in the base of the ladle
(b) The KALTEK liner is placed in position.
(c) The gap between the KALTEK lining and the shell is filled with coarse (10–20 AFS) sand
(d) A rammable material such as sodium silicate bonded sand, adequately vented, seals the top
(e) The FOSCAST refractory dam board is fitted and sealed

Figure 11.2 *Panel ladle made using KALTEK boards.*

Comparison of KALTEK cold ladle lining system with conventional ladle lining

A ductile iron foundry making automotive castings in the weight range 1 kg to 40 kg (average 5 kg) used pouring teapot ladles of 1100 kg capacity lined with high quality refractory concrete poured around an internal former, dried and fired with gas burners. Ladles were cleaned and repaired daily, each ladle had a lifetime of about 500 tonnes of liquid iron. The pouring practice was

Time $t = 0$ | Tap 1100 kg of liquid iron from the furnace to the treatment ladle
Time $t = 2$ minutes | Transfer to the pouring ladle with inoculation into the metal stream
Time $t = 5$ minutes | Begin pouring
Time $t = 16$ minutes | End pouring and coat the ladle lip with graphite wash
Back to $t = 0$ | Repeat cycle, two ladles are in use simultaneously

The cycle remained the same when the foundry changed from using concrete liners to using the KALTEK cold disposable liner system. For trials, a panel ladle was made using KALTEK boards (Fig. 11.2). This was later replaced by a shaped KALTEK two-piece lining (Fig. 11.3) with capacity 1100 kg. The new ladle lining consisted of three elements: a bottom bowl, a top ring and a concrete dam plate. Installation of the lining involved:

The bottom bowl was laid on a coarse sand bed
A refractory glue joint was deposited on the horizontal joint
The top ring was installed on the glue joint
The gap between the KALTEK lining and the shell was filled with coarse (15 AFS) sand
The top of the sand was sealed with a rammable material to prevent sand leakage when the ladle was tilted. The rammed material was vented by 1 mm holes, 50 mm apart

Figure 11.3   *1100 kg capacity teapot KALTEK ladle.*

After several years of experience with the KALTEK ladles a comparison of the two lining processes showed:

Furnace temperature reduced by 20°C with the same average pouring temperature
Less wear of the furnace refractory lining
Lower variance of pouring temperature
Lower ladle shell external temperature, better working environment
Lower direct labour costs on lining operations
Less gas consumption for preheat
Savings in FeSiMg alloy additions because of the lower tap temperature
Significant reduction in non-metallic inclusions
Better process control

KALTEK insulating lining system for automatic pouring boxes

Many automatic pouring systems, such as the DISAMATIC system, use refractory lined pouring boxes using a clay-graphite stopper rod and graphite nozzle. The box has a capacity of 1800 kg with an average fill of 600–900 kg. The conventional pouring box is lined with a 70% alumina castable having a life of about 25 shifts with a certain amount of patching needed daily. The nozzle requires maintenance every $1-1^1/_2$ hours to remove slag and metal build-up.

The KALTEK insulating board system consists of three components:

A back-up refractory safety liner in the metal shell of the pouring box
KALTEK boards custom-designed to fit the safety liner
Coarse sand to separate the KALTEK boards from the safety liner

The stopper rod, nozzle and pouring procedure remain the same.

The back-up 50% alumina liner is cast into the metal shell, since it has no direct contact with the molten metal, it lasts for years without maintenance. The KALTEK boards are supplied as custom-fitted sides, ends and bottom (Fig. 11.4). Metal banding is used to secure the assembled board set until placement in the pouring box. Before the boards are placed into the box, the graphite nozzle is set into the nozzle well and secured with a graphitic, thermally setting, ramming mix.

25 mm of coarse sand is spread evenly across the box floor and the pre-assembled KALTEK set lowered into the box (Fig. 11.5). More sand is poured into the cavity between side boards and safety liner. A refractory lid is attached and any exposed sand capped with a thin layer of plastic ram, vented with small holes to vent gases arising from organic binders in the boards. Finally the nozzle is finished according to standard practice.

The lining requires no heating before use, a small gas torch is used to

Figure 11.4 *Custom fitted KALTEK boards for DISAMATIC automatic pouring system.*

Figure 11.5 *Lowering the pre-assembled KALTEK boards into the DISAMATIC automatic pouring box.*

preheat the stopper rod and nozzle prior to the first cast. Pouring and casting are carried out as standard.

Benefits of the KALTEK lining system, in a ductile iron foundry, are:

30°C lower holding temperature
Alloy usage reduced because of lower holding temperature
Rate of heat loss one-third as fast
Refractory costs lowered
Labour costs lowered
Reduction of slag defects, misruns and shrinkage defects
Gas costs drastically lowered
Zero nozzle maintenance

Casting defects due to poor ladle maintenance

Slag and dross inclusions

These arise from slag carried over from the melting furnaces, from slag deposits left in the ladles after pouring and from fusion of the ladle lining material. They can be reduced by:

removal of adhering slag before use;
with tea-pot ladles, molten slag should be removed by back-tilting, to
avoid contamination of the spout;
use of good quality ganister and other lining materials.

Hydrogen pinholing

Caused by molten metal being poured from improperly dried ladles or
damp furnace launders. Hydrogen from the moisture is dissolved in the
liquid metal and released during solidification. The problem is increased by
trace amounts of aluminium in the metal. The defect is avoided by correct
ladle drying and pre-heating.
    The use of the KALTEK ladle lining system avoids the above problems.

Adding alloy additions in the ladle

Silicon and molybdenum additions can be made to cast iron by adding
LADELLOY fluxed additives.
    SILICON LADELLOY 33 is a fluxed ladle addition allowing the silicon
content of the iron to be increased quickly and accurately. The amount of
silicon to be added is determined by a wedge/chill test. SILICON LADELLOY
33 contains about 62% Si with a flux to ensure optimum pick-up. The required
quantity of additive is slowly poured into the metal stream as it passes
down the launder to the ladle. The use of ladle additions simplifies melting
operations since a standard charge can be adjusted at the ladle to cover a
wide variety of requirements.
    MOLYBDENUM LADELLOY 33 contains about 66%Mo and a small
amount of flux to ensure optimum pick-up of the element. The required
number of units should be placed in the bottom of the ladle and metal
tapped onto them. The phosphorus content of the iron must be less than
0.2% to avoid the formation of molybdenum phosphide.

Molten metal handling in steel foundries

Ladle practice

There are three types of ladle used for steel casting: lip pour, teapot and
bottom pour.

Lip pour ladles

In this type of ladle the metal is discharged over the lip, flow is controlled by tilting the ladle using a geared handwheel (Fig. 11.6a). Since the metal flows from the top of the ladle, the metal surface must be slag-free or a skimmer must be used to prevent slag entering the mould. The advantages of lip pouring ladles are that there are no narrow passages in which the metal can freeze so they are ideal for handling small quantities of metal. They are used for pouring small steel castings. They are inexpensive easy to prepare and are immediately ready for use. The disadvantage is that slag can easily be entrained in the metal stream.

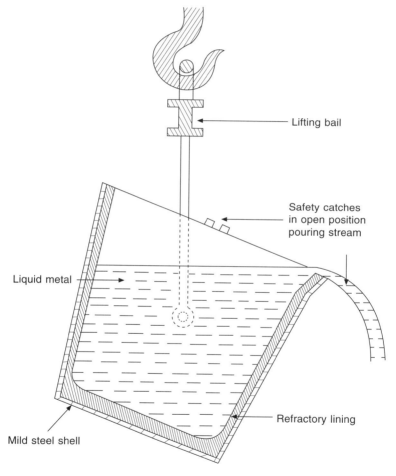

Figure 11.6a *Section through a lip pour ladle. (From Jackson, W.J. and Hubbard, M.W.,* Steelmaking for Steelfounders, *1979, SCRATA. Courtesy CDC.)*

Teapot ladles

A refractory dam before the ladle lip ensures that metal is drawn from the bottom of the ladle so that the stream is slag free (Fig. 11.6b). Deoxidation products have more time to float away from the area where metal is being withdrawn so the steel is generally cleaner than from a lip pour ladle. The disadvantage is that the narrow 'spout' may occasionally permit the liquid steel to freeze if the heat is tapped cold or pouring is prolonged.

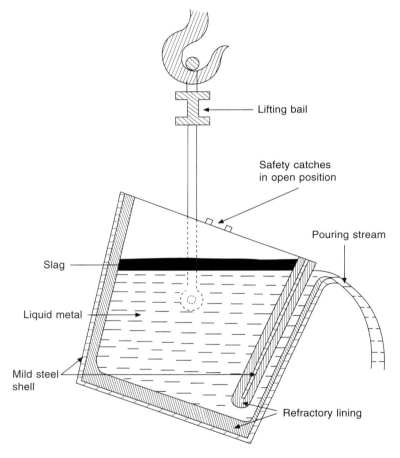

Figure 11.6b   *Section through a teapot ladle. (From Jackson, W.J. and Hubbard, M.W.,* Steelmaking for Steelfounders, *1979, SCRATA. Courtesy CDC.)*

With both lip pour and teapot ladles, it is only necessary to invert the ladle to remove all slag and metal before refilling or reheating.

Bottom pour ladles

The ladle is fitted with a pouring nozzle in its base, closed by a refractory stopper rod (Fig. 11.6c). The metal is drawn from the bottom and is therefore slag-free and non-metallics such as deoxidation products are able to float out of the melt. The metal stream flows vertically downwards from the ladle so that there is no movement of the stream during pouring. The disadvantage is that the velocity and rate of flow change during pouring as the ferrostatic head changes.

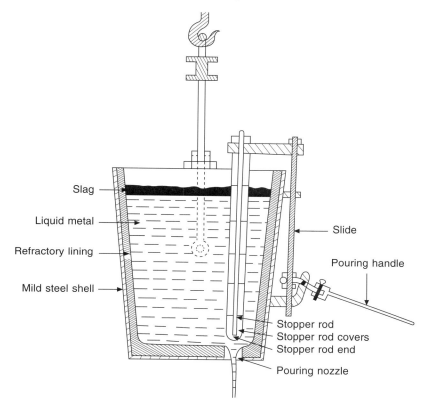

Figure 11.6c   *Section through a bottom pour ladle. (From Jackson, W.J. and Hubbard, M.W.*, Steelmaking for Steelfounders, *1979, SCRATA. Courtesy CDC.)*

Unless a reusable system is fitted, the nozzle and stopper rod assembly must be changed after each use, thereby increasing the turn-round time, costs and the number of ladles required to handle a given output of metal. The stopper rod may also distort or erode making it impossible to shut off the stream completely. It is not practical to handle less than 100 kg of steel due to the chilling effect of the stopper rod assembly. The flow from a bottom pour ladle depends on the size of the nozzle and the height of metal in the ladle so the flow rate and velocity of the metal stream reduces as the ladle empties. The nozzle/stopper rod is an excellent on-off valve but is not an effective flow control valve. Attempts to use it to control flow result in breakup of the metal stream with consequent risk of reoxidation of the steel. However it is possible to calculate the discharge rate and metal velocity for each ladle and nozzle and its variation throughout the pour so that ladles can be suited to the mould sizes that are being cast (see Chapter 18).

Ladle linings

The ideal lining is highly refractory, non-reactive with metal or slag and of low thermal conductivity and heat capacity. Fireclay is the traditionally used material either in the form of pre-fired bricks or as a plastic ramming grade. Brick linings must be installed by skilled personnel to ensure tight joints. Monolithic refractories are easier to install and eliminate the weak spots associated with bricks.

Although fireclay is the lowest cost refractory, it is possible to raise the level of as-cast quality considerably by using better lining materials. High alumina monolithic linings are popular because of their better refractoriness and their longer life.

Whether bricks, rammed monolithic refractories or castables are used, there is a long preparation and drying time needed. New ladle linings must be dried before use to remove moisture. After preliminary air drying, gas heaters are used to heat the lining to 600–800°C. Final firing of the lining occurs during its first use.

Steel rapidly loses temperature when held in ladles. A 5-tonne bottom pouring ladle lined with fireclay will lose 20°C in 5 minutes, while the same ladle lined with high alumina refractory will lose as much as 50°C in 5 minutes.

The KALTEK ladle lining system

The KALTEK range of products are disposable, insulating, refractory ladle linings for steel, iron and non-ferrous applications. The low thermal mass and insulating properties of KALTEK linings eliminate the need for pre-heat in virtually all ladle sizes and with most alloys. KALTEK inner linings can be quickly stripped and replaced when necessary.

KALTEK one-piece ladle linings are used for lip pouring in a wide range of ladle sizes. A FOSCAST castable alumina dam board can be integrated with the KALTEK liner to create a teapot ladle. Larger ladles, up to 15 tonnes, are lined with KALTEK in the form of board segments and bottom boards designed to suit individual ladles. KALTEK boards and linings are supplied in silica based refractory and also alumino-silicate for severe applications and high magnesia for special alloys.

The ladle to be lined with a pre-formed one piece lining must be free from holes. The KALTEK one-piece lining is fitted inside the shell with a minimum of 15 mm backfill of coarse, dry silica sand or BAKFIL, a coarse graded aggregate (AFS 35). The top exposed ring between lining and shell should be capped with a suitable mixture such as silicate bonded core sand and adequately vented with holes every 10–12 cm formed using 6 mm wire (Fig. 11.1).

Larger ladles to be lined with KALTEK boards must first be stripped and then furnished with a permanent base lining of alumino silicate bricks or castable refractory to act as a safety lining. The permanent lining must be dried and fired carefully. Vent holes in the ladles must be cleaned or new holes bored where necessary of 6–8 mm diameter at 20 cm centres.

The KALTEK segments and base boards are then carefully fitted inside the permanent lining, the joints between them being filled with KALSEAL refractory cement. The gap between the refractory and backing brick safety lining should be filled with BAKFIL or coarse, dry silica sand (Fig. 11.2). Refractory nozzle assemblies can be fitted into bottom pouring KALTEK lined ladles using special KALPACK ramming material (Fig. 11.7). The nozzle must be pre-heated to dry the KALPACK product before use.

The four methods available for a bottom pour ladle are shown here:

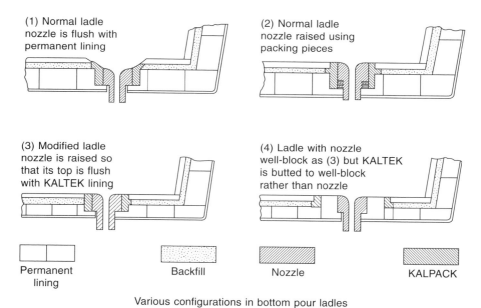

(1) Normal ladle nozzle is flush with permanent lining

(2) Normal ladle nozzle raised using packing pieces

(3) Modified ladle nozzle is raised so that its top is flush with KALTEK lining

(4) Ladle with nozzle well-block as (3) but KALTEK is butted to well-block rather than nozzle

Permanent lining   Backfill   Nozzle   KALPACK

Various configurations in bottom pour ladles

Figure 11.7 *Methods of installing nozzles in bottom pour ladles.*

KALTEK ladle linings are used under cold start conditions, pre-heating is unnecessary and would destroy the binder components of the boards. The thermal capacity of a KALTEK lined ladle is lower than conventional refractories and the thermal insulation twice as good so that while there is some initial chill when steel is tapped into the cold ladle, the superior insulation properties soon compensate so that lower tapping temperatures are possible (Fig. 11.8).

Initially, the concept of the KALTEK cold ladle lining system for steel foundries was based on single use of the disposable lining, but now in most cases, multiple life is possible. Pot ladles can be used many times as long as they are not allowed to cool between fillings. In many cases, bottom pour ladles can be used more than once as long as multi-life stopper rods such as the Roto-rod isostatically pressed one-piece alumina-graphite rod are used.

With KALTEK there is:

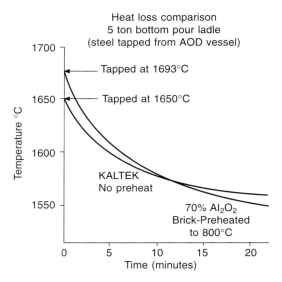

Figure 11.8   *Heat loss comparison, 5 ton bottom pour ladle, KALTEK v. brick-lined.*

No pre-heat
Faster ladle turnaround
Easier ladle maintenance
Better temperature control
Better working environment
Lower inclusion levels

Pouring temperature for steels

The temperature at which steel castings are poured is at least 50°C above
the liquidus temperature. Further superheat is needed to allow for cooling

Table 11.1   Variation of liquidus temperature with carbon content for Fe-C alloys

Carbon (%)	Liquidus temperature (°C)	Carbon (%)	Liquidus temperature (°C)
0.05	1533	0.55	1490
0.10	1528	0.60	1486
0.15	1524	0.65	1483
0.20	1520	0.70	1480
0.25	1515	0.75	1477
0.30	1511	0.80	1473
0.35	1507	0.85	1470
0.40	1502	0.90	1466
0.45	1498	0.95	1463
0.50	1494	1.00	1459

Table 11.2   Depression of liquidus temperature caused by the presence of 0.01% of alloying elements

Element	*Depression* (°C)	*Element*	*Depression* (°C)
P	0.300	Mo	0.020
S	0.250	Si	0.080
Mn	0.050	Cu	0.050
Cr	0.015	Sn	0.080
Ni	0.040	V	0.030

Example: a steel containing
0.06C, 1.0Si, 1.2Mn, 0.03P, 0.02S, 18Cr, 2.0Mo, 10.5Ni would have a liquidus temperature of
1532 − (8.0 + 6.0 + 0.90 + 0.50 + 27 + 4 + 42) = 1532 − 88.4 = 1444°C
The pouring temperature should be at least 1444 + 50 = 1494° say 1500°C

in the ladle during casting, which can be as high as 10°C per minute in high alumina lined ladles though much lower for KALTEK lined ladles.

The liquidus temperature of a steel can be estimated from Tables 11.1 and 11.2.

Chapter 12
Sands and green sand

Silica sand

Most sand moulds and cores are based on silica sand since it is the most readily available and lowest cost moulding material. Other sands are used for special applications where higher refractoriness, higher thermal conductivity or lower thermal expansion is needed.

Properties of silica sand for foundry use

Chemical purity

SiO_2	95–96% minimum	The higher the silica the more refractory the sand
Loss on ignition	0.5% max	Represents organic impurities
Fe_2O_3	0.3% max	Iron oxide reduces the refractoriness
CaO	0.2% max	Raises the acid demand value
K_2O, Na_2O	0.5% max	Reduces refractoriness
Acid demand value to pH_4	6 ml max	High acid demand adversely affects acid catalysed binders

Size distribution

The size distribution of the sand affects the quality of the castings. Coarse grained sands allow metal penetration into moulds and cores giving poor surface finish to the castings. Fine grained sands yield better surface finish but need higher binder content and the low permeability may cause gas defects in castings. Most foundry sands fall within the following size range:

Grain fineness number	50–60 AFS	Yields good surface finish at low binder levels
Average grain size	220–250 microns	
Fines content, below 200 mesh	2% max	Allows low binder level to be used
Clay content, below 20 microns	0.5% max	Allows low binder levels
Size spread	95% on 4 or 5 screens	Gives good packing and resistance to expansion defects
Specific surface area	120–140 cm^2/g	Allows low binder levels
Dry permeability	100–150	reduces gas defects

Grain shape

Grain shape is defined in terms of angularity and sphericity. Sand grains vary from well rounded to rounded, sub-rounded, sub-angular, angular and very angular. Within each angularity band, grains may have high, medium or low sphericity. The angularity of sand is estimated by visual examination with a low power microscope and comparing with published charts (Fig. 12.1).

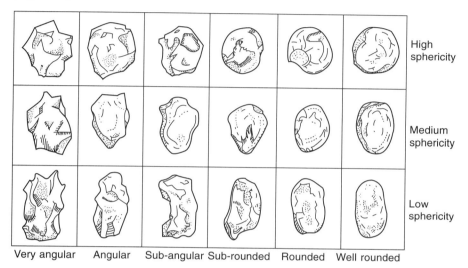

High
sphericity

Medium
sphericity

Low
sphericity

Very angular   Angular   Sub-angular Sub-rounded   Rounded   Well rounded

Figure 12.1   *Classification of grain shapes.*

The best foundry sands have grains which are rounded with medium to high sphericity giving good flowability and permeability with high strength at low binder additions. More angular and lower sphericity sands require higher binder additions, have lower packing density and poorer flowability.

Acid demand

The chemical composition of the sand affects the acid demand value which has an important effect on the catalyst requirements of cold-setting acid-catalysed binders. Sands containing alkaline minerals and particularly significant amounts of sea-shell, will absorb acid catalyst. Sands with acid demand values greater than about 6 ml require high acid catalyst levels, sands with acid demand greater than 10–15 ml are not suitable for acid catalysed binder systems.

Typical silica foundry sand properties

Chelford 60 Sand (a sand commonly used in the UK as a base for green sand and for resin bonded moulds and cores)

Grain shape: rounded, medium sphericity
Bulk density, loose: 1490 kg/m^3 (93 lb/ft^3)
GF specific surface area: 140 cm^2/g
Angle of repose: 33°

Chemical analysis:

SiO$_2$	Fe$_2$O$_3$	Al$_2$O$_3$	K$_2$O	Na$_2$O	CaO	TiO$_2$	Cr$_2$O$_3$	LOI
97.1	0.11	1.60	0.73	0.15	0.10	0.06	15 ppm	0.3

Acid demand (number of ml 0.1N HCl):

to:	pH3	pH4	pH5	pH6	pH7
ml:	2.0	1.8	1.4	1.0	0.8

Sieve grading of Chelford 60 sand

Aperture size (μm)	BSS mesh No.	% wt retained
1000	16	nil
700	22	0.4
500	30	2.3
355	44	10.0
250	60	25.7
210	72	23.8
150	100	28.7
105	150	7.6
75	200	1.3
−75	−200	0.2

AFS Grain Fineness No. 59
Base permeability: 106

Table 12.1 gives size gradings of typical foundry sands used in the UK and Germany.

Safe handling of silica sand

Fine silica sand (below 5 microns) can give rise to respiratory troubles. Modern foundry sands are washed to remove the dangerous size fractions and do not present a hazard as delivered. It must be recognised, however, that certain foundry operations such as shot blasting, grinding of sand covered castings or sand reclamation can degrade the sand grains, producing a fine quartz dust having particle size in the harmful range below 5 microns. Operators must be protected by the use of adequate ventilation and the wearing of suitable face masks.

Table 12.1 Typical UK and German foundry sands

Sieve size		Sand type				
		UK sands		German sands		
microns	*BSS No.*	*Chelford 50*	*Chelford 60*	*H32*	*H33*	*F32*
1000	16	trace	nil			
700	22	0.7	0.4			
500	30	4.5	2.3	1.0	0.5	1.0
355	44	19.8	10.0	15.0	7.5	7.0
250	60	44.6	25.7	44.0	30.0	30.0
210	72	21.6	23.8	39.0	60.0	60.0
150	100	8.2	28.7			
100	150	2.6	7.6			
75	200	nil	1.3	1.0	2.0	2.0
75	−200	nil	0.2	nil	nil	nil
AFS grain fineness No.		46	59	51	57	57
Average grain size		0.275 mm	0.23	0.27	0.23	0.23

Note: Haltern 32, 33 and Frechen 32 are commonly used, high quality German sands.
German sieve gradings are based on ISO sieves.
The German sands have rounder grains and are distributed on fewer sieves than UK sands, they require significantly less binder to achieve the required core strength.

Segregation of sand

Segregation, causing variation of grain size, can occur during sand transport or storage and can give rise to problems in the foundry. The greatest likelihood of segregation is within storage hoppers, but the use of correctly designed hoppers will alleviate the problem.

1. Hoppers should have minimum cross-sectional area compared to height.
2. The included angle of the discharge cone should be steep, 60–75°.
3. The discharge aperture should be as large as possible.

Measurement of sand properties

Acid demand value

Acid demand is the number of ml of 0.1 M HCl required to neutralise the alkali content of 50 g of sand.

Weigh 50 g of dry sand into a 250 ml beaker
Add 50 ml of distilled water
Add 50 ml of standard 0.1 M hydrochloric acid by pipette
Stir for 5 minutes
Allow to stand for 1 hour

Titrate with a standard solution of 0.1 M sodium hydroxide to pH values
of 3, 4, 5, 6 and 7
Subtract the titration values from the original volume of HCl (50 ml) to
obtain the Acid Demand Value

Grain size

See Section I, pp. 15, 16 for the method of measuring average grain size and
AFS grain fineness number.

Thermal characteristics of silica sand

Silica sand has a number of disadvantages as a moulding or coremaking
material.

It has a high thermal expansion rate (Fig. 12.2) which can cause expansion
defects in castings, such as finning or veining and scabbing.
It has a relatively low refractoriness (Table 12.2) which can cause sand
burn-on, particularly with steel or heavy section iron castings.
It is chemically reactive to certain alloys; for example, ferrous alloys
containing manganese. The oxides of Mn and Fe react with silica to form
low melting point silicates, leading to serious sand burn-on defects.

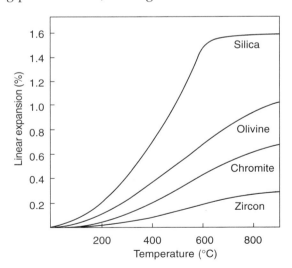

Figure 12.2   *Thermal expansion characteristics of zircon, chromite and olivine
sands compared with silica sand. (Courtesy CDC.)*

For certain types of casting, it may be necessary to use a non-silica sand,
even though all other sands are more expensive than silica.

Table 12.2   Sintering points of silica sand

Sand	Sintering point (°C)
High purity silica sand, >99% quartz	1450
Medium purity silica sand, 96% quartz	1250
Sea sand (high shell content)	1200
Natural clay bonded sand	1050–1150

Non-silica sands (Table 12.3)

Table 12.3   Properties of non-silica sands (compared with silica)

Property	Silica	Zircon	Chromite	Olivine
AFS grain size No.	60	102	74	65
Grain shape	rounded	rounded	angular	angular
Specific gravity	2.65	4.66	4.52	3.3
Bulk density (kg/m^3)	1490	2770	2670	1700
(lb/ft^3)	93	173	167	106
Thermal expansion 20–1200°C	1.9% non-linear	0.45%	0.6%	1.1%
Application	general	refractoriness chill	resistance to metal penetration chill	Mn steel

Zircon, ZrSiO$_4$

Zircon sand has a high specific gravity (4.6) and high thermal conductivity which together cause castings to cool faster than silica sand. The chilling effect of zircon sand can be used to produce favourable thermal gradients that promote directional solidification giving sounder castings. The thermal expansion coefficient of zircon is very low (Fig. 12.2) so that expansion defects can be eliminated. Zircon has higher refractoriness than silica, moreover it does not react with iron oxide, so sand burn-on defects can be avoided. Zircon sand generally has a fine grading, with AFS number between 140 and 65 (average grain size 115–230 microns), the most frequently used grade is around AFS 100.

Zircon is probably the most widely used of the non-silica sands. It is used with chemical binders for high quality steel castings and for critical iron castings such as hydraulic spool valves which contain complex cores, almost totally enclosed by metal, making core removal after casting difficult. Zircon has low acid demand value and can be used with all chemical binder systems. The Cosworth casting process uses the low thermal expansion of zircon sand cores and moulds to cast dimensionally accurate castings. The high cost of zircon sand makes reclamation necessary and thermal reclamation of resin bonded moulds and cores is frequently practised.

Zircon sands contain low levels of naturally occurring radioactive materials, such as uranium and thorium. Any employer who undertakes work with zircon mineral products is required, by law, to restrict exposure of workers to such naturally occurring contaminants so far as reasonably practical. The primary requirement is to prevent the inhalation of zircon dust. Suitable precautions are set out in the UK in HSE Guidance Note EH55: Dust – general principles of protection. Guidance on the possible hazards associated with the use, handling, processing, storage, transport or waste disposal of such naturally occurring radioactive materials and the control measures that are recommended to minimise exposure should be obtained from the supplier.

Chromite, $FeCr_2O_4$

The high specific gravity (4.5) and high thermal conductivity of chromite provide a pronounced chilling effect. Thermal expansion is low so expansion defects are unlikely to occur. Chromite sand has a glossy black appearance, it has greater resistance to metal penetration than zircon in spite of its generally coarser grading (typically AFS 70). It has somewhat higher acid demand than other sands which entails greater additions of acid catalyst when furane resin is used. Apart from this the sand is compatible with all the usual binder systems. Chromite is generally used for steel casting to provide chilling. It is difficult to reclaim chromite sand since, if it becomes contaminated with silica, its refractoriness is seriously reduced.

Olivine, Mg_2SiO_4

Olivine sand is used mainly for the production of austenitic manganese steel castings (which react with silica and other sands to give serious burn-on defects). It has also been used to avoid the health hazards possible with silica sand. Olivine has a very high acid demand and is not suitable for use with acid catalysed binders such as furan resins, it tends to accelerate the curing of phenolic urethane binders. Being a crushed rock, it is highly angular and consequently requires high binder additions. Thermal expansion is regular and quite low.

Green sand

The earliest method of bonding sand grains to form a sand mould was to use clay and water as a binder. The moulds could be used in the 'green' or undried state (hence the term green sand moulding) or they could be baked in a low temperature oven to dry and strengthen them to allow heavy castings to be made. Nowadays, dried, clay bonded sand is little used, having been replaced by chemically bonded sand, but green sand is still the most widely used moulding medium, particularly for iron castings. 40 years ago, 90% of all steel castings were produced in green sand moulds but this has now declined and the majority of steel castings are now made in chemically bonded sand moulds. This reflects the declining tonnage of high volume repetition steel castings now being made.

Originally, naturally occurring clay-sand mixtures were used, containing 10% or more of clay. It was found that by adding coal dust, the ease of stripping iron castings from the mould and the surface finish of the castings could be greatly improved. The heating of the coal dust by the liquid iron causes the formation of a type of carbon called lustrous carbon which is not wetted by the liquid iron, so the cast surface is improved. Coal dust additives are not used for steel casting since they would cause unacceptable carbon pick-up on the casting surface.

Clay bonded moulding sand can be used over and over again by adding water to replace that which is lost during casting, and re-milling the sand. However, clay which is heated to a high temperature becomes 'dead', that is it loses its bonding power. The coal dust is partly turned to ash by heat, so new clay, coal dust and water must be added and the sand re-milled to restore its bonding properties. As the sand is re-used, dead clay and coal ash build up in the sand, reducing its permeability to gases so that eventually water vapour and other mould gases are unable to escape from the mould and defective castings are produced.

Natural sands are little used nowadays (except for some aluminium castings) and most iron foundry green sands are 'synthetic' mixtures based on washed silica sand to which controlled additions of special moulding clays (bentonites) and low ash coal dust are made. Alternatively, special blended additives such as BENTOKOL or CARSIN, which combine the clay binder with lustrous carbon formers, can be used. Steel foundry green sands contain bentonites, starches and dextrins. The moulding sand becomes a 'sand system' which is continuously recycled, with suitable additions and withdrawals made as required. The control of system sand to achieve constant moulding properties has become an important technology of its own for it determines the quality of the castings produced.

Green sand additives

Base sand: Silica sand of AFS grain size 50–60 is usually used. The particle size distribution is important, a sand spread over 3 to 5 consecutive sieve sizes with more than 10% on each sieve gives the best results (see Table 12.1). Rounded or sub-angular sands are the best since the more round the grains, the better the flowability and permeability of the sand. The base sand should have the same size grading as the core sand used in the foundry, so that burnt core sand entering the system does not alter the overall size grading.

Clay: The best bonding clays are bentonites which can either have a calcium or a sodium base. Sodium bentonites occur naturally in the USA as Wyoming or Western bentonite and also in other countries particularly in the Mediterranean area. Green sand produced with this clay has medium green strength and high dry strength which increases the resistance to erosion of metal but can give problems at shakeout. Sodium bentonites are commonly used for steel casting production.

Calcium bentonites are more widely distributed, they produce green sands with rather high green strength but low dry strength so they have low erosion resistance and are prone to scabbing and other expansion defects. Calcium bentonites can be converted to sodium bentonite by adding soda ash, the calcium base is replaced by sodium and the clay then has properties approaching those of natural sodium bentonite. Such clays are known as activated clays.

Blended or mixed bentonites are commercial blends of sodium bentonite with calcium bentonite or a sodium-activated bentonite. Compositions can be varied to suit particular applications. Most iron foundries use blended bentonites.

Clays absorb water from the atmosphere so they should be stored in dry conditions, they do not deteriorate on storage, even for long periods.

Coal dust: Used mainly in iron foundries with some being used in non-ferrous foundries. The formation of 'lustrous carbon' from the thermal degradation of the volatiles given off during casting improves casting surface finish and strip. Good coal dust has the following properties:

Volatile content:	33–36%
Ash:	less than 5%
Fixed carbon:	50–54%
Sulphur:	less than 1%
Chlorine:	less than 0.03%
Size:	75 or 100 grade (AFS grain fineness No.)

It is important that the size grading of the coal dust used should not be too fine. Coal dust increases the moisture requirement of the sand and the finer it is, the more moisture is needed, which may have harmful effects on the castings. Fine coal dust will also reduce the overall permeability of the sand. Coal dust levels in green sand vary from 2 or 3% for small castings to 7 or 8% for heavy section castings. Too much coal dust can give rise to gas holes in the castings or misruns.

Coal dust must be stored dry to prevent the risk of fire. Damp coal dust can ignite spontaneously. Large stocks are undesirable and stocks should be rotated; first in, first out.

Coal dust replacements: Consist of blends of high volatile, high lustrous carbon materials blended with clays. They are generally more environmentally acceptable than coal dust, producing less fume during casting. They can be designed for particular applications such as high pressure moulding.

BENTOKOL and CARSIN additives are blends of natural clays to which have been added specially selected essential volatiles to provide both bond and volatile content for iron foundry green sand. They are in powder form, virtually dust free with flow properties and bulk densities similar to clay, they are supplied in bulk. Use of BENTOKOL/CARSIN often means that a 'one shot' sand addition is sufficient and always improves working conditions. Fume after casting is reduced and the foundry is cleaner and pleasanter while casting finish is maintained or enhanced.

Many of the BENTOKOL/CARSIN formulations contain a blend of natural sodium bentonite which has high burn-out temperature and results in lower concentrations of dead clay. This improves permeability, reduces moisture requirements and gives better all-round sand properties over a longer life cycle. Sodium bentonites take longer than other clays to develop their bond. To offset this BENTOKOL/CARSIN formulations may include dispersion agents so that the clay is rapidly spread and the bond quickly developed. BENTOKOL/CARSIN is compatible with sand systems based on clay and coal dust. Certain parameters of the sand system may change with the establishment of BENTOKOL/CARSIN; there may be a slight reduction of volatiles, LOI and clay fraction and the optimum moisture content will fall by up to 0.5%.

BENTOKOL 80 is a high volatile:clay ratio sand conditioner for the production of iron castings where exceptional surface finish is required. It is suitable for use in high pressure or boxless moulding lines. It is a free flowing grey powder. Because of its high volatile content, one part of BENTOKOL 80 can usually replace 1.8 parts of coal dust and will additionally replace a proportion of the clay addition. The use of BENTOKOL 80 ensures a positive and rapid response to addition rate changes so allowing immediate alteration of sand properties to match changes in production rate or pattern mix.

LUCIN is a coal dust replacement product having no bentonite.

Cereal binders: Used mainly in steel foundries to increase strength and toughness of the green sand. There are two main types of cereal binder: starch and dextrin. Starch is the basic material and is produced from a number of plant materials with maize starch being the most commonly used for foundry purposes. Starch is treated by heating to form 'pre-gelatinised starch' for foundry use. In this form it develops a gel when mixed with water at about 10% concentration. Dextrin is produced by treating starch with a weak acid and cooking at 120–150°C. The starch is re-polymerised and converted to dextrin which is freely soluble in water, forming syrups.

Starch and dextrin additions affect the bonding and moulding properties of green sand mixtures (Table 12.4).

The main benefits to moulding properties obtained by adding a cereal binder are:

Providing greater toughness to the mix and, therefore, the compacted

Table 12.4   The effect of starch and dextrin on green sand mixes

Property	Starch addition	Dextrin addition
Green strength	Increases moderately	Increases slightly
Toughness	Increases moderately	Increases markedly
Green deformation	Increases	Increases
Pattern stripping	Improves	Improves markedly
Mould compaction	Little effect	Can improve compaction with high pressure moulding
Sensitivity to change in moisture	Lowered	Lowered
Dry strength	Increases slightly	Increases markedly at higher moisture content
Friability of mould surfaces	Increases slightly	Markedly reduces

From SCRATA Technical Bulletin No. 16, Additives to Green Sand.

sand can deform more easily without fracture, i.e. the cereals correct the brittleness of the sand. This enables the sand (mould) to strip from difficult pockets in the pattern without failing. Starch and dextrin are both effective in this way.

Providing harder and less friable mould surfaces and edges where green sand moulds are allowed to air dry. Starch is not effective for this purpose but dextrin, being water soluble migrates to the mould surface forming a hard, less friable crust. Friable mould surfaces can cause dirty moulds and sand inclusions in the casting and dextrin corrects friability problems.

Effect of cereal binders on casting surface quality

Cereal additions do not improve the erosion resistance of the sand or resistance to metal penetration. They are very good anti-scabbing agents and the higher the amount of starch or dextrin the greater is the resistance to expansion scabbing. Starch is more effective in preventing expansion scabbing than a dextrin and should be employed when making larger green sand moulds and where radiation effects from the rising steel are most severe.

Use of cereals in moulding mixes

Effective additions of cereal binders to new green sand facing mixes are between 0.5 and 0.75%. Accurate weighing of the cereal addition is a primary requirement as is efficient mulling to develop the full properties of the relatively tough mixes.

In unit type green sands part of the cereal is destroyed during the casting process. Cereal binders lose their beneficial properties when heated beyond

225°C. A recommended addition on each recycle is 0.1 to 0.25% depending upon the amount of cereal burn-out and the dilution by cores and new sand.

Choice of the type of cereal binder to use depends upon the intended purpose for the green sand in the foundry. Dextrin is to be preferred for general green sand work and where moulds air dry before pouring. Starch is best for larger work where expansion scabbing is possibly a more serious problem. As a compromise and to give the beneficial effects from both types, a percentage of starch and dextrin should be added to the moulding mixes. A suitable proportion is two-thirds starch to one-third dextrin.

Water: Water is needed to develop the clay bond but it can cause casting defects. Where there is strong, localised heating, e.g. in the vicinity of ingates or on horizontal mould surfaces exposed to radiant heat from the metal, moisture is driven back from the mould surface, condensing in a wet, weak underlying layer that can easily fracture to produce expansion defects in castings such as scabs, rat-tails and buckles.

MIXAD additive: MIXAD 61 additive has been developed to eliminate sand expansion defects. By improving the wet strength of the condensed water layer, MIXAD 61 eliminates the associated defects. MIXAD 61 is added early in the milling cycle, usually 0.5–1.0% is used in the facing sand. If MIXAD 61 is used in a unit sand system, an addition of 0.05–0.10% is usually sufficient. MIXAD 61 is used principally in sands for iron and copper base alloys.

The green sand system

The 'sand system' in green sand foundries is illustrated in Fig. 12.3. It comprises the following items of equipment;

The bulk sand storage hopper
Returned knockout sand is stored in a hopper which ideally has capacity for about 4 hours of usage. Returned sand is very variable in properties, since some will have been heated very strongly by the hot metal while sand which came from the edges of the mould will have been heated very little. Some mixing of the sand occurs in the storage hopper. It is better to have several smaller volume hoppers rather than one large unit since mixing of the return sand is improved. Hoppers should be designed in such a way that sand flows evenly through without sticking to the walls.

The sand mill
At the mill, the additions of new sand, clay, coaldust and water are made to the returned sand. Sand mills may be continuous or batch but batch mills are nowadays preferred because better sand control is possible. Several

designs of mill are available, their purpose is to mix the sand and spread the moistened clay over the surface of the sand grains in order to develop the bond. Older sand mills used heavy vertically set mulling wheels and plough blades (Fig. 12.4). A certain optimum milling time is necessary and this has to be determined for each green sand system and adhered to carefully. Short milling times underuse the additives and produce sand having low and variable green strength. Modern sand mills use intensive mixers to develop the clay bond quickly and efficiently. Whatever type of mill is used, regular maintenance is essential to ensure that the milling efficiency does not change.

Additions of new sand, clay and coal dust (or coal dust replacement) are made at the mill. Batch mills use weighed additions while continuous mills use calibrated feeders to make continuous additions. Water is added, often via an automatic moisture controller which measures the moisture content of the sand as it feeds into the mixer batch hopper prior to mixing then calculates the exact amount of water to be added to the sand batch discharged into the mixer to bring the moisture to the required level. Prepared sand is usually passed through an aerator which 'fluffs up' the sand before feeding the moulding machines.

Buffer hopper
A buffer hopper is usually provided before the moulding machines, so that a moulding machine stoppage does not necessitate running freshly milled sand straight back to the main hopper.

The moulding machine(s)
These can be jolt-squeeze, high pressure, impact moulding machines etc. Each has its advantages and each requires special moulding sand properties. See p. 164.

Casting
The moulds are cast then allowed to cool for a suitable time, often 30–40 minutes, before knocking out the castings. The cooling time not only allows the castings to solidify completely but also reduces internal stresses within the casting caused by differential cooling of sections of varying thickness.

Shake-out
The shakeout separates the castings from the sand. It may be a vibrating grid or a rotating drum, the latter also having a cooling function since water is sprayed onto the sand in the drum. The shakeout is provided with copious air extraction, to prevent dust from entering the foundry. This also removes some fines from the sand and so plays an important part in the control of the sand system.

Return sand conveyor
Sand from the shakeout contains lumps of partially burnt cores, very hot burnt core sand, hot moulding sand from near the casting and cooler moulding sand from the edges of moulds. A magnetic separator removes metallic

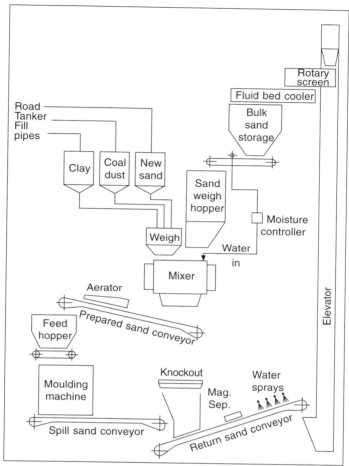

Figure 12.3   *Flow diagram for a typical green sand plant. (From* Foundryman, *March 1998, p. 102. Courtesy Foundry and Technical Liaison Ltd.)*

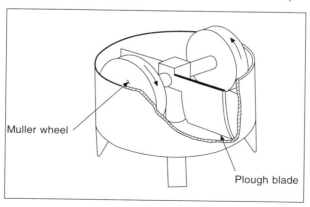

Figure 12.4   *Vertical wheel batch muller. (Sixth Report of Institute Working Group T30, Mould and Core Production.* Foundryman, *Feb. 1986.)*

particles and water sprays are used to effect some preliminary cooling before elevation, since very hot sand can damage the elevator.

Screen
The return sand is elevated to a screen where residual core lumps and other unwanted material are removed before the sand is returned to the hopper. Non-magnetic metallics are also removed by the screen.

Sand cooling
Hot sand causes problems of excessive moisture loss and condensation of moisture on patterns and cores, so the returned sand must be cooled. The only effective way of cooling green sand is by evaporation of water. Evaporation of 1% water cools the sand by about 25°C, so water is sprayed onto the sand and air drawn over the sand to evaporate it, either in the rotary drum shakeout or at the screen or in a specially designed fluidised bed cooler. Returned green sand cannot be fluidised by air alone. Sand is delivered onto a screen which is vibrated to transport the sand along it. Water is added by sprays and evaporated by blowing air through the perforated screen.

Sand removal or addition
The sand system must have a facility for removal of excess sand, since foundries making highly cored castings build up the sand level in the system through the introduction of burnt core sand. Regular addition of new sand is needed to maintain control of green sand properties.

Green sand properties

Green sand for iron foundries typically has the following properties:

	jolt/squeeze machines	*high pressure (DISA etc.)*
water	3–4%	2.5–3.2%
green strength	70–100 kPa	150–200 kPa
	10–15 psi	22–30 psi
compactability	45–52%	38–40%
permeability	80–110	80–100
live clay	5.0–5.5%	6.0–10.0%
volatiles	2.5%	2.0%
LOI	7.0–7.5%	6.0%

Steel foundries use sand having similar properties except for reduced volatiles and LOI since coal dust is not used.

A typical grading of a sand suitable for iron or steel castings is:

Sieve mesh size (μm)	%
710	0.14
500	1.42
355	5.44
250	22.88
212	19.64
150	26.74
106	9.70
75	2.84
pan	1.10

Control of green sand systems

When iron or steel is poured into green sand moulds, the heat from the metal drives off some water from the sand and clay, burns some of the coal dust to coke and ash (in the case of iron), and burns a proportion of the clay so that its bonding properties are destroyed. Before the sand can be used again its properties must be restored by removing burnt clay, coal dust and ash, and milling in new clay, coal dust (or dextrin) and water. The following additions are usually necessary at the sand mill in order to maintain the required moulding properties of the sand.

> 0.3–0.5% of new clay
> 0.3–0.5% of coal dust (or 0.2% dextrin)
> 1.5–2.5% of water

In addition, it is desirable to add a proportion (up to 10%) of new silica sand to the system and to dispose of a corresponding amount of used system sand. If highly cored work is made, the amount of new sand addition can possibly be reduced since well-burnt core sand may serve as new sand, but care must be taken since certain core binder residues (particularly from phenolic-isocyanate binders) can harm the green sand properties.

The precise amount of the additions that must be made depends on many factors, including:

> The weight and type of castings being made, which affects the burn-out
> The amount of core sand entering the system
> Whether hold-ups on the moulding line have allowed freshly milled sand to return directly to the sand hopper

Good sand control depends on carefully monitored experience of the particular sand system, since every system is different. The normal additions of clay, coal dust or BENTOKOL/CARSIN at each cycle represent only about 10% of the total active clay or coal present in the system. It is not possible to change the total clay or coal dust content quickly since any change in the addition rate takes about 20 cycles to work its way fully into the system. For example; an addition of 0.3% clay is usually sufficient to maintain the total clay level at 3.0%. If the clay addition is increased to 0.4%, the total clay content after one cycle will only rise to about 3.1% and it will take 20 cycles (about 1 week) for the full effect of the change of addition to be felt and for the clay level to rise to around 4.0%.

The operator must find by experience what additions are needed during normal operation of the foundry to maintain the required physical properties of the sand; changes will normally only be necessary if some change of practice has occurred, such as a change to heavier castings, which will cause more burn-out of clay and coal. A change to more highly cored castings is likely to result in more burnt core sand entering the system, which will require greater additions to compensate.

Because sand systems have such built-in inertia, or resistance to change, the only way that quick changes can be made to the moulding properties of the sand is by increasing or reducing the addition of water to the sand mill. This has an immediate effect on sand properties and as long as the other constituents of the sand are in control, the sand's mouldability can quickly be corrected.

Sand testing

Regular testing of the properties of the sand is essential. One or two sand tests do not truly indicate the condition of the whole sand system, since a sand sample weighs only about 1 kg and cannot represent the whole 200 tonnes or so of the system. At least five samples should be taken per shift and measured for moisture, green strength, compactability and permeability. LOI and volatiles should be measured once per day. Active clay, twice per week.

Records of additions should also be kept:

> Weight of clay, coal dust and new sand added each day
> Number of moulds made
> Weight of iron poured
> Weight of used sand removed from the system each day

The three-ram test method developed by the AFS in the 1920s is still widely used to measure green strength and compactability although it is not really suitable for sand used in modern high pressure moulding machines. The AFS ramming test is being replaced by tests on 50 mm diameter specimens produced by squeeze machines which can reproduce the pressures actually developed on the mould by the machine that the foundry uses.

Control graphs

Individual figures mean rather little, but daily average sand properties should be plotted together with weekly figures of active clay and average additions of clay, coal dust and new sand. After a few weeks of plotting the data, it will be possible to draw control lines. Variation within the lines is permissible but if results appear outside the control lines, then action must be taken such as increasing or reducing clay or coal or new sand. If action is taken, it must be remembered that it will be 1 or 2 days before the full effect of the action will be seen.

The sand properties, moisture, green strength, compactability and permeability are of first importance and must be maintained. The remaining graphs are of secondary importance, but provide useful information which may allow problems to be anticipated. It is also necessary to keep a log book for the sand system listing information such as:

When the sand mill was maintained
When new sand was added and old sand removed
Changes in foundry practice, such as change of core binder, or change in type of casting made.

Parting agents

Unless a water-repellent material is present on the pattern surface to prevent wetting, green sand will tend to stick to the pattern when rammed. This will cause damage to the mould when the pattern is withdrawn. To prevent this, it is usual to spray the pattern with a liquid parting agent such as one of the SEPAROL grades. The frequency of application depends on the type and quality of the pattern, but every 10th or 12th mould is typical.

Special moulding materials, LUTRON

LUTRON moulding sand is a waterless sand containing mineral oils designed for use with fine sands to produce a superfine finish in castings such as name plates, plaques, medallions and other art castings made in aluminium or copper base alloys.

The ready-mixed LUTRON moulding sand is based on a very fine sand (AFS No. 160). The sand has excellent flowability due to the lubricating properties of the bond. It can be used alone or as a facing sand backed with normal green sand or other type of sand. If used as a unit sand, the mix can be reconstituted after casting by adding a compensating amount of LUTRON binder and milling.

The LUTRON binder can be used with any suitable fine, dry sand; additions of 10–12% are normally needed.

Green sand moulding machines

The moulding machine must compact the green sand evenly around the pattern to give the mould sufficient strength to resist erosion while liquid metal is poured, to withstand the ferrostatic pressure exerted on the mould when full and to resist the expansion forces that occur on solidification of the iron. Green sand does not move easily under compression forces alone so that achieving uniform mould strength over a pattern of complex shape by simply squeezing is not possible. Various combinations of jolting and vibrating with squeeze applied simultaneously or sequentially have been used to produce uniform strength moulds at up to 400 moulds per hour. Green sand moulds have traditionally been made in steel flasks, but flaskless moulding is now widely used for smaller castings.

Moulding in flasks

The basic principles used are:

Jolt-Squeeze A pattern plate carrying the pattern surrounded by a moulding flask is fitted onto a jolt piston. Green sand from a hopper above the machine fills the moulding box loosely. The assembly is raised by the pneumatic jolt cylinder and allowed to fall against a stop. The jolt action is repeated a pre-set number of times causing the sand to be compacted to some extent, the density being highest nearest the pattern plate. Additional compaction is then provided by a pneumatic or hydraulic squeeze plate (Fig. 12.5a). More uniform squeeze can be achieved by using compensating squeeze heads (Fig. 12.5b) which compact the sand more uniformly even though the depth of sand over the pattern varies.

Vibration-squeeze Instead of a jolt table, the pattern and flask are vibrated before and during squeezing by a multi-ram head.

Shoot-squeeze A shoot head above the flask contains loose green sand. Opening a shoot valve allows compressed air stored in a reservoir to travel through the shoot head causing a column of sand to be shot into the flask where the kinetic energy of the rapidly moving sand compacts it against the pattern. A hydraulic cylinder then squeezes the mould to complete the compaction.

Impulse compaction The flask is filled loosely with prepared sand which is compacted by means of a pressure wave generated by explosion or by compressed air. Compaction is greatest near the mould surface. Squeeze pressure may then be applied to compress the back of the mould. Deep pattern draws are possible with this method.

Vacuum squeeze A partial vacuum is created in the flask surrounding the

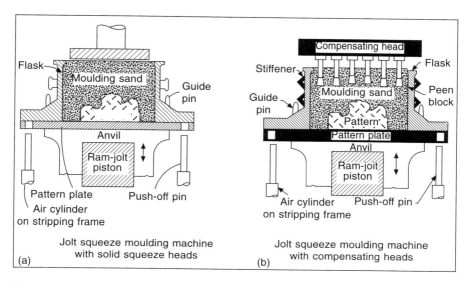

Figure 12.5 *Jolt squeeze moulding machine:* (a) *with solid squeeze head;* (b) *with compensating heads. (Sixth Report of Institute Working Group T30, Mould and Core Production.* Foundryman, *Feb. 1996.)*

pattern. A metered amount of sand is released into the chamber where the vacuum accelerates the sand which impacts onto the pattern causing compaction. A multi-ram head provides high pressure squeeze to complete the compaction of the mould. The system is suitable for large moulds.

Flaskless moulding

Horizontally parted (match-plate moulding)

A matchplate is a pattern plate with patterns for both cope and drag mounted on opposite faces of the plate. Both cope and drag halves of the mould are filled with prepared sand in the machine before being brought together for the high pressure squeeze with simultaneous vibration to compact the sand. The completed mould is pushed out of the machine onto a shuttle conveyor. Moulds can be made at up to 200 per hour.

Vertically parted moulding

The Disamatic flaskless moulding machine introduced in the late 1960s (now supplied by Georg Fischer Disa) revolutionised green sand moulding, allowing high precision moulds to be made at up to 350 moulds/hour. The method of operation is shown in Fig. 12.6. One pattern half is fitted onto the end of a hydraulically operated squeeze piston with the other pattern half fitted to a swing plate, so called because of its ability to move and swing

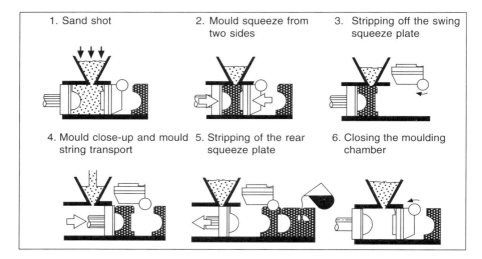

1. Sand shot

2. Mould squeeze from two sides

3. Stripping off the swing squeeze plate

4. Mould close-up and mould string transport

5. Stripping of the rear squeeze plate

6. Closing the moulding chamber

Figure 12.6   *Vertically parted flaskless moulding, the Disamatic machine. (Sixth Report of Institute Working Group T30, Mould and Core Production.* Foundryman, *Feb. 1996.)*

away from the completed mould. Sand from a supply hopper above the machine is blown into the moulding chamber by means of a variable pressure compressed air supply stored in a nearby air receiver. Vacuum can be applied to the moulding chamber to vent air and assist in drawing sand into deep pattern recesses. Both halves of the pattern are hydraulically squeezed together to compress the sand block. As the swing plate moves away, the piston pushes the new mould to join ones previously made, to form a continuous mould string. Mould sizes available are from 500 mm × 400 mm × 315 mm on the smallest 2110 model, up to 950 mm × 800 mm × 635 mm on the largest model manufactured, the 2070. Flexibility is available through variable mould output, variable mould thickness, fast pattern change and core placing options. Varying degrees of control sophistication are provided dependent on the model. Cores can be placed in the mould using a mechanised core placer.

There are many variations on the moulding principles described above. See Sixth Report of Institute of British Foundrymen Working Group T30 (*Foundryman*, Feb. 1996, p. 3) from which some of the above information has been taken.

Chapter 13
Resin bonded sand

Chemical binders

A wide variety of chemical binders is available for making sand moulds and cores. They are mostly based either on organic resins or sodium silicate (see Chapter 14), although there are other inorganic binders such as cement, which was the earliest of the chemical binders to be used; ethyl silicate, which is used in the Shaw Process and for investment casting and silica sol, which is also used for investment casting.

The binders can be used in two ways:

As self-hardening mixtures; sand, binder and a hardening chemical are mixed together; the binder and hardener start to react immediately, but sufficiently slowly to allow the sand to be formed into a mould or core which continues to harden further until strong enough to allow casting. The method is usually used for large moulds for jobbing work, although series production is also possible.

With triggered hardening; sand and binder are mixed and blown or rammed into a core box. Little or no hardening reaction occurs until triggered by applying heat or a catalyst gas. Hardening then takes place in seconds. The process is used for mass production of cores and in some cases, for moulds for smaller castings.

Self-hardening process (also known as self-set, no-bake or cold-setting process)

Clean, dry sand is mixed with binder and catalyst, usually in a continuous mixer. The mixed sand is vibrated or hand-rammed around the pattern or into a core box; binder and catalyst react, hardening the sand. When the mould or core has reached handleable strength (the strip time), it is removed from the pattern or core box and continues to harden until the chemical reaction is complete.

Since the binder and catalyst start to react as soon as they are mixed, the mixed sand has a limited 'work time' or 'bench life' during which the mould or core must be formed (Fig. 13.1). If the work time is exceeded, the final strength of the mould will be reduced. Work time is typically about one-third of the 'strip time' and can be adjusted by controlling the type of catalyst and its addition rate. The work time and strip time must be chosen to suit the type and size of the moulds and cores being made, the capacity of the sand mixer and the time allowable before the patterns are to be re-used. With some binder systems the reaction rate is low at first, then speeds up so that the work time/strip time ratio is high. This is advantageous, particularly for fast-setting systems, since it allows more time to form the mould or core.

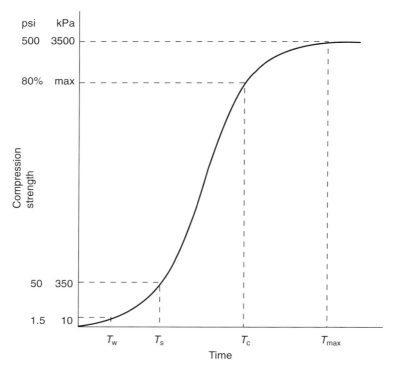

Figure 13.1   *Typical hardening curve for self-hardening sand:*
T_w = *work time*
T_s = *strip time*
T_c = *casting time*
T_{max} = *time to achieve maximum strength.*

Stripping is usually possible when the sand has reached a compression strength of around 350 kPa (50 psi) but the actual figure used in practice depends on the type of binder system used, the tendency of the binder to sag before it is fully hardened, the quality of the pattern equipment and the complexity of the moulds and cores being made.

It is advisable to strip patterns as soon as it is practical, since some binder chemicals attack core box materials and paints after prolonged contact. The properties of chemical binders can be expressed in terms of:

Work time (bench life): which can be conveniently defined as the time after mixing during which the sand mixture has a compressive strength less than 10 kPa, at this stage it is fully flowable and can be compacted easily.
Strip time: which can be defined as the time after mixing at which a compressive strength of 350 kPa is reached, at this value most moulds and cores can be stripped without damage or risk of distortion.
Maximum strength: the compressive strength developed in a fully hardened mixture, figures of 3000–5000 kPa are often achieved.

It is not necessary to wait until the maximum strength has been achieved before moulds can be cast, the time to allow depends on the particular castings being made, usually casting can take place when 80% of the maximum strength has been reached.

Testing chemically bonded self-hardening sands

Units

Compressive strength values may be reported in

SI units	$kPa = kN/m^2$
cgs units	kgf/cm^2
Imperial units	$psi = lbf/in^2$

Conversion factors:

$$100 \text{ kPa (kN/m}^2) = 1.0197 \text{ kgf/cm}^2$$
$$= 14.5038 \text{ psi (lbf/in}^2)$$
$$1 \text{ kgf/cm}^2 = 98.0665 \text{ kPa}$$
$$= 14.22 \text{ psi (lbf/in}^2)$$
$$1 \text{ psi (lbf/in}^2) = 6.895 \text{ kPa (kN/m}^2)$$
$$= 0.07032 \text{ kgf/cm}^2$$

Conversion table

kPa (kN/m^2)	kgf/cm^2	psi (lbf/in^2)
10	0.10	1.5
50	0.51	7.3
100	1.02	14.5
200	2.04	29.0
300	3.06	43.5
400	4.08	58.0
500	5.10	72.5
600	6.12	87.0
700	7.14	101.5

800	8.16	116.0
900	9.18	130.5
1000	10.20	145.0
2000	20.39	290.1
3000	30.59	435.1
4000	40.79	580.1
5000	50.99	725.2

The curing properties (work time, strip time and maximum strength) are measured by compression tests using 50 mm diameter specimen tubes with end cups, or AFS 2 inch diameter tubes, with a standard rammer. Sand is mixed in a food mixer or small coresand mixer; catalyst being added first and mixed, then the resin is added and mixed.

Measurement of 'work time' or 'bench life'

Mix the sand as above, when mixing is complete, start a stopwatch and discharge the sand into a plastic bucket and seal the lid.

After 5 minutes, prepare a standard compression test piece and immediately measure the compressive strength.

At further 5 minute intervals, again determine the compressive strength, stirring the mixed sand in the bucket before sampling it.

Plot a graph of time against strength and record the time at which the compressive strength reaches 10 kPa (0.1 kgf/cm^2, 1.5 psi); this is the work time or bench life.

The sand temperature should also be recorded.

For fast-setting mixtures, the strength should be measured at shorter intervals, say every 1 or 2 minutes.

Measurement of strip time

Prepare the sand mixture as before.

When mixing is complete, start a stop-watch.

Prepare 6–10 compression test pieces within 5 minutes of completion of mixing the sand.

Cover each specimen with a waxed paper cup to prevent drying.

Determine the compressive strength of each specimen at suitable intervals, say every 5 minutes.

Plot strength against time.

Record the time at which the strength reaches 350 kPa (3.6 kgf/cm^2, 50 psi), this is the 'strip time'.

The sand temperature should also be recorded.

Measurement of maximum strength

Prepare the sand mixture as before.

Record the time on completion of mixing.

Prepare 6–10 specimens as quickly as possible covering each with a waxed cup.

Determine the strength at suitable intervals say, 1, 2, 4, 6, 12, 24 hours.

Plot the results on a graph and read the maximum strength.

The sand temperature should be held constant if possible during the test.

While compressive strength is the easiest property of self-hardening sand to measure, transverse strength or tensile strength are being used more frequently nowadays, particularly for the measurement of maximum strength.

Mixers

Self-hardening sand is usually prepared in a continuous mixer, which consists of a trough or tube containing a mixing screw. Dry sand is metered into the trough at one end through an adjustable sand gate. Liquid catalyst and binder are pumped from storage tanks or drums by metering pumps and introduced through nozzles into the mixing trough; the catalyst nozzle first then binder (so that the binder is not exposed to a high concentration of catalyst).

Calibration of mixers

Regular calibration is essential to ensure consistent mould and core quality and the efficient use of expensive binders. Sand flow and chemical flow rates should be checked at least once per week, and calibration data recorded in a book for reference.

Sand: Switch off the binder and catalyst pumps and empty sand from the trough. Weigh a suitable sand container, e.g. a plastic bin holding about 50 kg. Run the mixer with sand alone, running the sand to waste until a steady flow is achieved. Move the mixer head over the weighed container and start a stop watch. After a suitable time, at least 20 seconds, move the mixer head back to the waste bin and stop the watch. Calculate the flow in kg/min. Repeat three times and average. Adjust the sand gate to give the required flow and repeat the calibration.

Binders: Switch off the sand flow and the pumps except the one to be measured. Disconnect the binder feed pipe at the inlet to the trough, ensuring that the pipe is full. Using a clean container, preferably a polythene measuring jug, weigh the binder throughput for a given time (minimum 20 seconds). Repeat for different settings of the pump speed regulator. Draw a graph of pump setting against flow in kg/min.

Repeat for each binder or catalyst, taking care to use separate clean containers for each liquid. Do not mix binder and catalyst together, since they may react violently. Always assume that binders and catalysts are hazardous, wear gloves, goggles and protective clothing.

When measuring liquid flow rate, the pipe outlet should be at the same height as the inlet nozzle of the mixer trough, so that the pump is working against the same pressure head as in normal operation.

Mixers should be cleaned regularly. The use of STRIPCOTE AL applied to the mixer blades, reduces sand build-up.

Sand quality

In all self-hardening processes, the sand quality determines the amount of binder needed to achieve good strength. To reduce additions and therefore cost, use high quality sand having:

AFS 45–60 (average grain size 250–300 microns)
Low acid demand value, less than 6 ml for acid catalysed systems
Rounded grains for low binder additions and flowability
Low fines for low binder additions
Size distribution, spread over 3–5 sieves for good packing, low metal penetration and good casting surface.

Pattern equipment

Wooden patterns and core boxes are frequently used for short-run work. Epoxy or other resin patterns are common and metal equipment, usually aluminium, may be used for longer running work. The chemical binders used may be acid or alkaline or may contain organic solvents which can attack the patterns or paints. STRIPCOTE AL aluminium-pigmented suspension release agent or silicone wax polishes are usually applied to patterns and core boxes to improve the strip of the mould or core. Care must be taken to avoid damage to the working surfaces of patterns and regular cleaning is advisable to prevent sand sticking.

Curing temperature

The optimum curing temperature for most binder systems is 20–25°C but temperatures between 15 and 30°C are usually workable. Low temperatures retard the curing reaction and cause stripping problems, particularly if metal pattern equipment is used. High sand temperatures cause reduction of work time and poor sand flowability and also increase the problem of fumes from the mixed sand. If sand temperatures regularly fall below 15°C, the use of a sand heater should be considered.

Design of moulds using self-hardening sand

Moulds may be made in flasks or flaskless. Use of a steel flask is common for large castings of one tonne or more, since it increases the security of casting. For smaller castings, below one tonne, flaskless moulds are common. Typical mould designs are illustrated in Fig. 13.2. The special features of self-hardening sand moulds are:

Large draft angle (3–5°) on mould walls for easy stripping
Incorporation of a method of handling moulds for roll-over and closing
Means of location of cope and drag moulds to avoid mismatch
Reinforcement of large moulds with steel bars or frames
Clamping devices to restrain the metallostatic casting forces
Use of a separate pouring bush to reduce sand usage
Mould vents to allow gas release
Sealing the mould halves to prevent metal breakout
Weighting of moulds if clamps are not used
Use of minimum sand to metal ratio to reduce sand usage, 3 or 4 to 1 is typical for ferrous castings

Foundry layout

With self-hardening sand; moulds and cores are often made using the same binder system, so that one mixer and production line can be used. A typical layout using a stationary continuous mixer is shown in Fig. 13.3. The moulds may or may not be in flasks. Patterns and core boxes circulate on a simple roller track around the mixer. The length of the track is made sufficient to allow the required setting time, then moulds and cores are stripped and the patterns returned for re-use.

For very large moulds, a mobile mixer may be used.

Sand reclamation

The high cost of new silica sand and the growing cost of disposal of used foundry sand, make the reclamation and re-use of self-hardening sands a matter of increasing importance. Reclamation of sand is easiest when only one type of chemical binder is used. If more than one binder is used, care must be taken to ensure that the binder systems are compatible. Two types of reclamation are commonly used, mechanical attrition and thermal.

Wet reclamation has been used for silicate bonded sand. The sand is crushed to grain size, water washed using mechanical agitation to wash off the silicate residues, then dried. The process further requires expensive water treatment to permit safe disposal of the wash water so its use is not common.

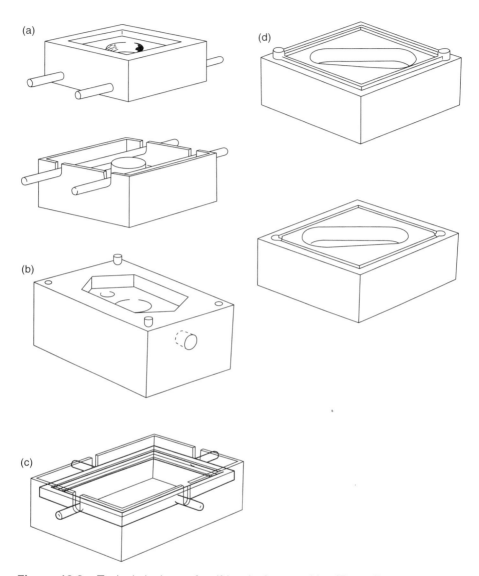

Figure 13.2 *Typical designs of self-hardening moulds. (From Foundry Practice Today and Tomorrow, SCRATA Conference, 1975.) (a) Method of moulding-in steel tubes for ease of handling boxless moulds. (b) Sockets moulded into boxless moulds for ease of lifting, roll-over and closing. (c) Steel reinforcement frames for handling large boxless moulds. (d) Method of locating mould halves and preventing runout.*

The difficulty and cost of disposing safely of used chemically bonded sand has led to the growing use of a combination of mechanical and thermal treatment. Mechanical attrition is used to remove most of the spent binder. Depending on the binder system used, 60–80% of the mechanically reclaimed sand can be rebonded satisfactorily for moulding, with the addition of clean sand. The remaining 20–40% of the mechanically treated sand may then be

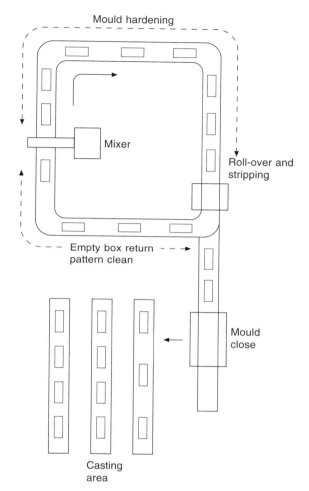

Mould hardening

Mixer

Roll-over and stripping

Empty box return pattern clean

Mould close

Casting area

Figure 13.3 *Foundry layout for self-hardening sand moulds.*

thermally treated to remove the residual organic binder, restoring the sand to a clean condition. This secondarily treated sand can be used to replace new sand. In some cases, all the used sand is thermally treated.

Mechanical attrition

This is the most commonly practised method because it has the lowest cost. The steps in the process are:

Lump breaking; large sand lumps must be reduced in size to allow the removal of metal etc.
Separation of metal from the sand by magnet or screen.
Disintegration of the sand lumps to grain size and mechanical scrubbing

to remove as much binder as possible, while avoiding breakage of grains.

Air classification to remove dust, fines and binder residue.

Cooling the sand to usable temperature.

Addition of new sand to make up losses and maintain the quality of the reclaimed sand.

Reclamation by attrition relies on the fact that the heat of the casting burns or chars the resin binder close to the metal. Even at some distance from the metal, the sand temperature rises enough to embrittle the resin bond. Crushing the sand to grain size followed by mechanical scrubbing then removes much of the embrittled or partially burnt binder. The more strongly the sand has been heated, the more effectively is the sand reclaimed.

Mechanical attrition does not remove all the residual binder from the sand, so that continued re-use of reclaimed sand results in residual binder levels increasing until a steady state is reached which is determined by:

the amount of burn-out which occurs during casting and cooling
the effectiveness of the reclamation equipment
the percentage of new sand added
the type of binder used.

The equilibrium level of residue left on the sand is approximately expressed as:

$$P = \frac{TB}{1 - TR}$$

P is the maximum percentage of resin that builds up in the sand (the LOI of the reclaimed sand)
B is the binder addition (%)
T is the fraction of binder remaining after reclamation
R is the fraction of sand re-used

Example: In a typical furane binder system

$B = 1.4\%$ resin + 0.6% catalyst = 2.0%

$T = 0.7$ (only 30% of the binder residue is removed)

$R = 0.90$ (90% of reclaimed sand is re-used with 10% new sand)

$$P = \frac{0.7 \times 2.0}{1 - (0.7 \times 0.9)}$$

= 3.78% (residual binder that builds up on the sand)

This represents an inefficient reclaimer. Ideally P should not exceed 3.0%. Even with an inefficient reclaimer $P = 3\%$ can be achieved by reducing R, that is, by adding more new sand. For example, reducing R to 0.75 (25% addition of new sand) reduces P to 2.95%

Regular testing of reclaimed sand for LOI, acid demand, grain size and temperature is needed, together with regular maintenance of the reclaimer to ensure that consistent mould quality is achieved.

Binder systems containing inorganic chemicals, e.g. silicate based systems, alkaline phenolic resins or binder systems containing phosphoric acid are difficult to reclaim at high percentages because no burn-out of the inorganic material occurs.

Use of reclaimed sand with high LOI may cause problems due to excessive fumes at the casting stage, particularly if sulphonic acid catalysed furane resins are used.

Thermal reclamation

Sand bonded with an entirely organic binder system can be 100% reclaimed by heating to about 800°C in an oxidising atmosphere to burn off the binder residues, then cooling and classifying the sand. Thermal reclaimers are usually gas heated but electric or oil heating can also be used. The steps in the process are:

Lump breaking
Metal removal
Heating to about 800°C for a certain time in a fluidised bed furnace or rotary kiln
Cooling the sand, using the extracted heat to preheat the incoming sand or the combustion air
Classification
Addition of new sand to make up losses in the system.

Thermal reclamation is costly because of the large amount of heat needed and the relatively expensive equipment. The ever growing cost of sand disposal, however, leads to its increasing use.

Sand losses

Whatever method of reclamation is used, there is always some loss of sand so that 100% reclamation can never be achieved. Sand losses include: burn-on, spillage, inefficiencies in the sand system and the need to remove fines. Dust losses of around 5% can be expected and total sand losses of up to 10% may be expected.

Typical usage of sand reclamation

Furane bonded sand

Mechanical attrition allows up to 90% of sand to be re-used. Only sulphonic acid catalysed sand can be reclaimed. Reclaimed sand may have up to 3% LOI. Binder additions on rebonding can be reduced by 0.15–0.2% (from say, 1.2% on new sand to 1.0%) with a proportionate reduction of catalyst. Nitrogen and sulphur levels build up and may cause casting defects and excessive fume if allowed to exceed 0.15% when the sand is rebonded. Low nitrogen resins must be used.

Thermal treatment at 800°C can be used on the residual sand. Nitrogen and sulphur are removed with the remaining organic content.

Phenolic urethane bonded sand

Up to 90% of sand can be reclaimed by mechanical attrition. If iron oxide is added to the sand mix, care must be taken to ensure that the iron content of the sand does not rise too high or the refractoriness may suffer.

Thermal reclamation can be used on the residual sand as long as the iron oxide content is low.

Cold box core sand, from core assembly casting processes, can be thermally reclaimed satisfactorily. If uncontaminated core lumps can be separated from green sand, they can also be reclaimed.

Alkaline phenolic bonded sand

These binders contain potassium which, if allowed to rise above 0.15% in the reclaimed sand, will cause unacceptable reduction of work time and final strength. This limits attrition reclamation to about 70%.

The residue after attrition reclamation can be thermally reclaimed if it is treated with FENOTEC ADT 1, an anti-fusion additive. This additive prevents sintering of the sand during thermal treatment and aids removal of potassium salts.

Resin shell sand

Shell moulding and core sand (silica or zircon) is fully reclaimable by thermal means. The used sand often has as much as 2% of residual resin left after casting. Useful heat can be obtained from the combustion of the resin residues which significantly reduces the quantity of heat that must be supplied. The quality of the reclaimed sand must be regularly checked for:

Size grading
LOI (the visual appearance of the sand, which should be the colour of new sand, gives a good indication of LOI)
Inorganic residues, e.g. calcium stearate additions, which are often made to lubricate shell sand, burn to CaO during reclamation, giving rise to high acid demand.

Foundries wishing to thermally reclaim shell sand must have equipment to re-coat the sand for re-use.

Silicate-ester bonded sand

Standard silicate-ester bonded sand can be reclaimed mechanically only at rather low levels, less than 50% because the build-up of soda in the sand reduces its refractoriness. The presence of reclaimed sand in the re-bonded mix reduces the work time and final strength and increases the tendency for sagging.

VELOSET special silicate ester binder (see Chapter 14)

Up to 90% reclaimed sand can be re-used with the VELOSET system. Shakeout sand is reduced to grain size in a vibratory crusher which provides the primary attrition stage. The sand is then dried in a fluidised bed drier. Secondary attrition takes place next in a hammer mill. The sand is finally passed through a cooler-classifier ready for re-use. The reclaimed sand is blended with new sand in the proportion 75 to 25. During the first 10 cycles of re-use, the sand system stabilises and the bench life of the sand increases by a factor of up to 2. Also, mould strength should improve, and it is usually possible to reduce the binder addition level by up to 20% yet still retaining the same strength as achieved using new sand. Once the process has become established, it becomes possible to re-use up to 85–90% of the sand (Figs 14.4, 14.5).

Wet reclamation

Some silicate bonded sands are particularly difficult to reclaim mechanically because Na_2O builds up and lowers refractoriness. (VELOSET sand is an exception.) Thermal reclamation is ineffective, but water washing can be used. The steps in the process are:

Lump breaking
Metal separation by magnet or screen
Disintegration to grain size
Water wash with mechanical agitation to wash off the silicate residues
Separation of sand and water
Drying of the sand
Agglomeration of the alkaline residues in the water to allow settlement and separation
Water treatment to permit safe disposal of the water

Wet reclamation is expensive and its use is not common.

Self-hardening resin binder systems

Furanes

Foseco products: FUROTEC, ESHANOL Binders.

Principle: Self-setting furane sands use a furane resin and an acid catalyst. The resins are urea-formaldehyde (UF), phenol-formaldehyde (PF), or UF-PF resins with additions of furfuryl alcohol (FA). Speed of setting is controlled by the percentage of acid catalyst used and the strength of the acid.

Sand: Since the resins are acid catalysed, sands should have a low acid demand, less than 6 ml.

Resin: A wide range of resins is available having different nitrogen content. The UF base resin contains about 17–18% N; furfuryl alcohol is N-free, so the N content of a UF-FA resin depends on its FA content. Nitrogen can cause defects in steel and high strength iron castings, so it is advisable to use high FA resins (80–95% FA, 3.5 – 1% N) although they are more expensive than lower FA resins. These resins are particularly useful with sands of high quality, such as the German Haltern and Frechen sands, which are round-grained and can be used with low resin additions (<1.0%) so the total nitrogen on the sand is low.

A range of UF-PF furanes is also available, having low nitrogen content, they are frequently used with sands of rather lower quality, which require quite high resin additions (>1.2%). The total nitrogen on the sand can be kept to a lower level than is possible with UF-FA resins.

Phenol-formaldehyde resin is N-free and a range of totally N-free PF-FA resins is available, these are used for particularly sensitive castings, or when sand is reclaimed. Where the presence of phenol in the resin presents a problem, due to restrictions on used sand disposal, special phenol-free FA-Resorcinol resins (FA-R resins) can be used. They contain less than 0.5%N and are used for steel and high quality iron casting. While they are expensive, they are used at low addition rates.

Catalyst: UF-FA resins can be catalysed with phosphoric acid, which has the advantage of low odour and no sulphur but reclamation of the sand is difficult. FA-R, PF-FA, UF-FA with high FA and UF-PF-FA resins are catalysed with sulphonic acids, either PTSA (para-toluene sulphonic acid), XSA (xylene sulphonic acid) or BSA (benzene sulphonic acid), the last two are strong acids and are used when faster setting is required. Sulphonic acid catalysed resin sand can be reclaimed but the sulphur content of the sand rises and may cause S defects in ferrous castings particularly in ductile iron. Environmental problems may also arise due to the SO_2 gas formed when moulds are cast. Mixed organic and inorganic acids may also be used.

Addition rate: Resin: 0.8–1.5% depending on sand quality. Catalyst: 40–60%

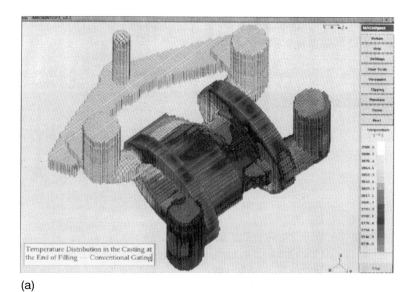

(a)

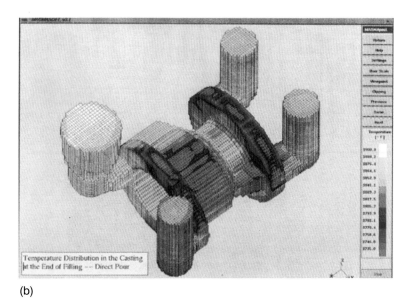

(b)

Figure 18.13 (a) Temperature distribution within a valve casting immediately after filling through a conventional gating system. Large areas are much colder than the last metal poured. (b) Temperature distribution within a valve casting immediately after filling through a KALPUR unit shows higher overall temperature, illustrating reduced heat loss when direct pour units are used.

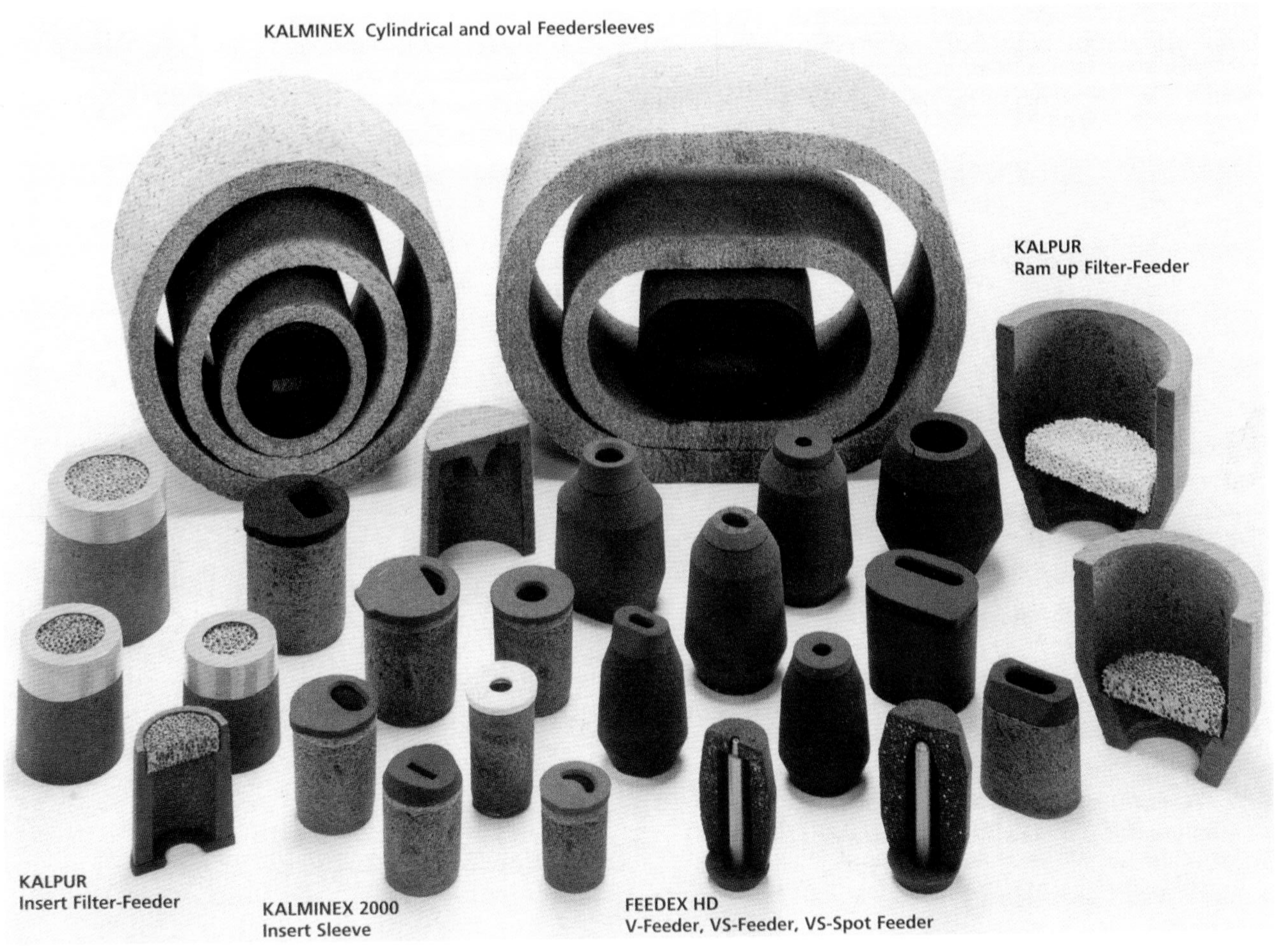

KALMINEX Cylindrical and oval Feedersleeves

KALPUR
Ram up Filter-Feeder

FEDEX HD
V-Feeder, VS-Feeder, VS-Spot Feeder

KALMINEX 2000
Insert Sleeve

KALPUR
Insert Filter-Feeder

Figure 19.27 Some of the available Foseco feeding and filter-feeding units.

Figure 20.1 Solid model of a valve casting utilising 40 million elements.

Figure 20.2 Solidification contours for the lower section of the valve casting.

Figure 20.3 An 'X-ray' plot showing predicted shrinkage cavities.

of resin, depending on sand temperature and speed of setting required. If fast setting is required, XSA or BSA catalyst should be used.

Pattern equipment: Wood or resin is commonly used. Metal patterns can be used but if the pattern is cold, setting may be retarded. FA is a powerful solvent and will attack most paints, patterns must be clean and preferably unpainted. Special release agents such as STRIPCOTE AL can be used.

Temperature: The optimum is 20–30°C. Cold sand seriously affects the setting speed and the final strength.

Strength: Depends on resin type and addition, typically 4000 kPa (600 psi) compression strength.

Speed of strip: Can be from 5–30 minutes. Short strip times require high speed mixers and may cause problems of sand build-up on the mixer blades.

Work time/strip time: UF-FA resins are better than PF-FA, the higher the FA content, the better.

Coatings: Water or spirit based coatings may be used, alcohol based coatings may cause softening of the mould or core surface and it is advisable to allow $^1/_2$–1 hour between stripping and coating.

Environmental: Formaldehyde is released from the mixed sand, so good extraction is needed around the mixer. When sulphonic acid catalyst is used, SO_2 gas is evolved when moulds are cast, particularly if reclaimed sand is used.

Reclamation: Attrition reclamation works well and it is often possible to reduce the resin addition on reclaimed sand. Low N resin and sulphonic acid catalyst must be used. The LOI of the reclaimed sand must be kept to less than 3.0%. N and S in the sand should not exceed 0.15% when the sand is rebonded.

Casting characteristics: N above 0.15% in the sand will cause defects in iron and steel castings. S above 0.15% in the sand may cause reversion to flake graphite at the surface of ductile iron castings. Metal penetration can be a problem if work time is exceeded.

General: Probably the most widely used self-hardening system because of the easy control of set times, good hot strength, erosion resistance and the ease of reclamation.

Phenolic-isocyanates (phenolic-urethanes)

Foseco product: POLISET Binder.

Principle: The binder is supplied in three parts. Part 1 is a phenolic resin in an organic solvent. Part 2 is MDI (methylene diphenyl diisocyanate). Part 3 is a liquid amine catalyst. When mixed with sand, the amine causes a reaction between resin and MDI to occur, forming urethane bonds which rapidly sets the mixture. The speed of setting is controlled by the type of catalyst supplied in Part 1.

Sand: The binder is expensive, so good quality sand is needed to keep the cost of additions down. AFS 50–60 is usually used.

Addition rate: The total addition is typically 0.8% Part 1, 0.5% Part 2, more or less being used, depending on the sand quality.

Pattern equipment: Wood, resin or metal can be used. Paints must be resistant to the strong solvents in Part 1 and Part 2.

Temperature: Sand temperature affects cure rate, but not as seriously as other binders. The optimum temperature is 25–30°C.

Speed of strip: Can be very fast, from 2–15 mins. A high speed mixer is needed to handle the fast cure rates.

Work-time/Strip-time: Very good.

Strength: Compression strength is typically over 4000 kPa (40 kgf/cm^2, 600 psi).

Coatings: If water-based coatings are used, they should be applied immediately after strip and dried at once. Spirit based coatings should be applied 15–20 minutes after strip and are preferably air-dried.

Casting characteristics: Lustrous carbon defects may occur on ferrous castings. Part 2 (MDI) contains 11.2% N and gas defects may occur in steel castings. The binder has rather low initial hot strength, so erosion defects may occur. All these problems can be reduced by the addition of 1 or 2% of iron oxide to the sand.

Reclamation: Attrition works well, 90% re-use of sand is possible. If iron oxide is added to the sand mix, care must be taken to ensure that the iron content of the sand does not rise too high or the refractoriness may suffer.

Environmental: Solvent release during mixing and compaction may be troublesome, use good exhaust ventilation. Isocyanates can cause respiratory problems in sensitised individuals. Once an individual has become sensitised, exposure, however small, may trigger a reaction. MDI has very low vapour pressure at ambient temperature, so exposure to the vapour is unlikely to cause problems, however uncured isocyanate may be present on airborne sand particles at the mixing station. Good exhaust ventilation is essential.

General: This system has never been as popular in the UK and Europe as in the USA due to the difficulty of stripping and the low hot strength. Steel foundries find it useful for cores, where the low hot strength reduces hot tearing.

Alkaline phenolic resin, ester hardened

Foseco products: FENOTEC, FENOTEC hardener.

Principle: The binder is a low viscosity, highly alkaline phenolic resole resin. The hardener is a liquid organic ester. Sand is mixed with hardener and resin, usually in a continuous mixer. The speed of setting is controlled by the type of ester used.

Sand: Can be used with a wide range of sands including zircon, chromite and high acid demand sand such as olivine.

Resin addition: 1.2–1.7% depending on the sand quality; 18–25% hardener based on resin.

Nitrogen content: Very small or zero.

Pattern equipment: Wood or resin, patterns strip well.

Temperature: Low sand temperature slows the cure rate, but special hardeners are available for cold and warm sand.

Speed of strip: 3 min to 2 hours depending on the grade of hardener used. The work time/strip time ratio is good.

Strength: 24 h strength:				
	transverse	1600 kPa	16 kgf/cm^2	230 psi
	tensile	900	9	130
	compression	4000	40	600

Coatings: Some iron and steel castings can be made without coatings. Both water based and spirit based coatings can be used, but some softening-back of the sand surface is possible. The stripped core or mould should be allowed to harden fully before applying the coating. Water based coatings should be dried as quickly as possible. Alcohol based coatings should be fired as soon as possible after application.

Casting characteristics: Good as-cast finish on all metals. Hot tearing and finning defects are eliminated. No N, S or P defects. Good breakdown, particularly on low melt point alloys. Widely used for steel castings as well as iron and aluminium.

Reclamation: FENOTEC binders allow up to 70–90% sand reclamation by attrition, there is some loss of strength and careful management of the alkali content of the reclaimed sand is needed. The remainder of the sand can be thermally reclaimed if it is treated with FENOTEC ADT 1, an anti-fusion additive which aids potassium removal.

Environmental: Low fume evolution at mixing, casting and knockout stages.

General: Stripping from all pattern types is excellent; this, together with the good casting finish achieved and the low fume evolution, makes the system popular for all types of casting but most of all for steel.

Triggered hardening systems

Cores for repetition foundries are usually made using a triggered hardening system with the mixed sand being blown into the core box. The cores must be cured in the box until sufficient strength has been achieved to allow stripping without damage or distortion, usually the core continues to harden after stripping. Transverse breaking strength or tensile strength are used to assess the properties of triggered core binder systems since they represent the properties needed to strip and handle cores. The strength requirements needed depend on the particular type of core being made. Thin section cores, such as cylinder head water jacket cores, require high stripping strength because of their fragile nature. Tensile strengths of 1000–2000 kPa (150–300 psi) are typical, equating roughly to transverse strengths of 1500–3000 kPa. Final strengths may be higher, but some binder systems are affected by storage conditions (humidity in particular) and strengths may even fall if the storage conditions are poor. Surface hardness is also important but difficult to measure precisely.

   Heavy section cores and moulds made by core blowing can be successfully made with lower strength binders, and factors such as fume released on casting and compatibility of core residues with the green sand system may be more important than achieving the very highest strength.

Heat triggered processes

The sand and binder are mixed then introduced into a heated core box or pattern. The heat activates the catalyst present in the binder system and cures the binder. Core boxes and patterns must be made of metal, normally grey cast iron, evenly heated by means of gas burners or electric heating elements. The working surface of the core box is usually heated to 250°C, higher temperatures may overcure or burn the binder. Sand is a rather poor conductor of heat and heat penetration is rather slow (Fig. 13.4), it is therefore difficult to cure thick sections of sand quickly and fully. When the core is ejected from the core box, residual heat in the sand continues to penetrate into the core, promoting deeper cure. To achieve the fastest cure times, heavy section cores are often made hollow, using heated mandrels or 'pull-backs' to reduce core thickness. Core boxes are usually treated with silicone release agents to improve the strip. The thermal expansion of the metal core box must be taken into account when designing core boxes. A cast iron core box cavity 100 mm long will expand by 0.27 mm when heated from 25 to 250°C. This change becomes significant on large cores.

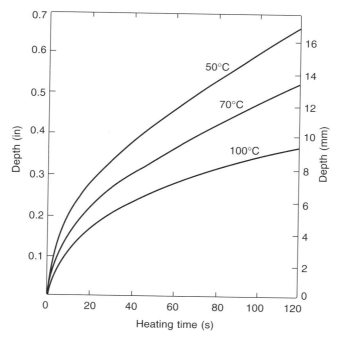

Figure 13.4 *Temperature rise in a sand core with one surface in contact with a heated core box at 250°C (theoretical).*

Other methods of applying heat to sand cores have been tried. Microwave or dielectric heating is difficult because electrically conducting metal core boxes cannot be used. Certain resins can be used for core boxes but they pick up heat from the cores and may distort.

Blowing heated air through the core has also been used, but a large volume of air is needed and the method becomes slow and impractical if core sections above about 30 mm are to be cured. Approximately 1 kg (800 litres at STP) of heated air is needed to heat 1 kg of sand to curing temperature.

Gas triggered systems

Sand and binder are mixed and blown into a core box then a reactive gas is blown into the core box causing hardening of the binder. Hardening occurs at room temperature, avoiding the need for heated core boxes. Many of the gases used to trigger chemical reactions are highly reactive and toxic so that specially designed gas generators are needed to meter the gas accurately into the core box. Core boxes must be sealed to contain the gases and often scrubbers must be used to absorb the exhaust gas to prevent it from contaminating the atmosphere both inside and outside the foundry. Frequently, the gassing cycle includes an air purge to remove residual gas from the core so that it can be handled and stored safely.

Gas cured binder systems often allow very fast curing, even of thick section cores.

Heat triggered processes

The Shell or Croning process

Principle: Sand is pre-coated with a solid phenolic novolak resin and a catalyst to form a dry, free-flowing material. The coated sand is blown into a heated core box or dumped onto a heated pattern plate causing the resin to melt and then harden. Cores may be solid or they may be hollow with the uncured sand shaken out of the centre of the core. Shell moulds are normally only 20–25 mm thick.

Sand: The base sand is usually a fine silica sand of AFS 60-95. To achieve the highest strength with the minimum resin addition, the sand should be pure, rounded grain and free from surface impurities. Zircon sand is also pre-coated for special applications.

Resin and catalyst additions: The resins are solid phenolic novolaks which melt at around 100°C. The catalyst is hexamine, a white powder. Resin additions are 2.5–4.5% (depending on the application and the strength required), hexamine is added at 11–14% of the resin content.

Nitrogen content: The resins are N-free but hexamine contains 40% N. Special low N systems with little or no hexamine are available.

Pre-coating procedure: There are two processes; hot coating and warm coating. In the hot coating procedure, the sand is heated to around 130°C and charged into a batch mixer. Solid resin granules are added, the heat from the sand melts the resin which then coats the sand grains. Aqueous hexamine solution plus release agents are added, cooling the sand below the melt point of the resin. The coated sand is broken down to grain size and finally cooled.

In warm coating, the sand is heated to 50°C and charged into a batch mixer together with hexamine and release agents. Resin in alcohol solution is added and warm air blown through the mixer to vaporise the solvent, leaving the sand evenly coated. After breaking down the sand lumps and cooling, the coated sand is ready for use.

Many foundries purchase ready coated sand from specialist suppliers.

Bench life: Pre-coated sands last indefinitely if stored dry and not exposed to excessive heat, which may cause clumping.

Core blowing: The free-flowing sand can be blown at low pressure, 250–350 kPa(35–50 psi). Core boxes should be made of cast iron heated to 250°C. Copper or brass must be avoided since ammonia released during curing will cause corrosion.

Curing times: Minimum curing time is 90 seconds but 2 minutes is common. The longer the curing time, the thicker is the shell build-up. Increasing curing temperature does little to speed up cure rate and runs the risk of overcuring the sand adjacent to the core box or pattern plate.

Core strength: Surface hardness and strength is high on ejection. A 3.5% resin content will give a tensile strength of 1400 kPa (200 psi). Storage properties of cured cores is excellent.

Gas evolution: 4–5 ml/g of gases for each 1% of resin.

Casting characteristics: Core and mould coatings are unnecessary. The surface finish of castings is excellent. Breakdown is good, particularly with hollow cores. Accuracy and reproduction of pattern detail is extremely good because of the free-flowing properties of pre-coated sand. Some distortion, particularly of moulds, can occur on casting and support of shell moulds by steel shot or loose sand may be needed for heavy section castings. N defects may occur on steel castings unless low-N coated sand is used. Excellent surface finish is achieved on aluminium castings.

Health hazards: Phenol and ammonia are released during curing so good ventilation is needed at core and mould making machines. Casting into shell moulds produces unpleasant and hazardous fumes and often shell moulds are cooled in an enclosed tunnel, allowing fumes to be collected and exhausted outside the foundry.

Reclamation: When both moulds and cores are made by shell process, the sand can be reclaimed thermally.

Shell moulding: Shell moulds are made by dumping pre-coated sand onto an iron pattern plate heated to 240–260°C. After a suitable time, usually about 2 minutes, the mould is overturned, returning the uncured sand to the hopper and leaving a shell mould 20–25 mm thick which is ejected from the pattern plate. Cores are placed and the two half-moulds are glued together with hot-melt adhesive (CORFIX).

Shell moulds are usually cast horizontally, unsupported. If cast vertically, support is usually necessary to prevent mould distortion. Steel shot is frequently used to support the mould.

KALMIN pouring cups are often used with horizontally poured shell moulds.

General: The shell process was one of the first synthetic resin foundry processes to be developed. Although it is slow and rather expensive, it is still widely used, because of the excellent surface finish and dimensional accuracy of the castings produced.

Hot box process

Principle: The binder is an aqueous PF-UF or UF-FA resin, the catalyst is an aqueous solution of ammonium salts, usually chloride and bromide. Sand is mixed with the liquid resin and catalyst and blown into a heated core box. The heat liberates acid vapour from the catalyst which triggers the hardening reaction. Hardening continues after removal of the core from the box causing the sand to be cured throughout.

Sand: Clean silica sand of AFS 50-60, acid demand should be low. The binder is aqueous so traces of moisture in the sand do not affect the cure. Low sand temperatures (below 20°C) slow the cure rate.

Resin addition: 2.0–2.5% depending on sand quality. Catalyst additions are 20–25% of the resin weight.

Nitrogen content: The resin contains 6–12%N. The catalysts are usually ammonia chloride solutions and also contain N. The total N in the system is 10–14% depending on the resin used. Low-N or N-free systems are available but their performance is not as good as the N-containing binders.

Mixing procedure: Batch or continuous mixers can be used, add catalyst first then the resin.

Bench life: 1–2 hours, mixed sand must be kept in closed containers to prevent drying out. High ambient temperatures reduce bench life.

Core blowing: High blowing pressure, 650–700 kPa (90–100 psi) is needed to blow the wet sand into core boxes.

Core boxes: Cast iron or steel, fitted with gas or electric heaters to maintain the temperature at 250–300°C. Silicone parting agents are usually applied to improve the strip of the core.

Curing time: Thin section cores cure in 5–10 seconds. As cores increase in section, curing time must be extended up to about one minute for 50 mm section. For larger sections still, it is advisable to lighten-out cores with heated mandrels, otherwise the centre of the core will remain uncured.

Core strength: Surface hardness and strength is high on ejection. Final tensile strength is 1400–2800 kPa (200–400 psi). Storage properties are good if cores are kept dry.

Casting characteristics
Ferrous castings: Cores are usually coated to prevent burn-on. Water based coatings, such as RHEOTEC are used. The hot strength and breakdown is good, particularly for the UF-PF-FA resins, the higher the PF content, the worse the breakdown. The rather high nitrogen can cause gas holes and fissures in iron castings and additions of 1–3% of a coarse grained form of iron oxide are often used to minimise N-defects. Finely powdered red iron oxide can also be used but it causes some loss of strength.
   Hot box cores made using phenolic resin expand slightly during casting and this can cause distortion, for example on large water jacket cores. Iron oxide additions reduce the expansion, so does the use of UF/FA resins.
   Aluminium castings: Special hot box resins are available which break down readily at aluminium casting temperatures.
   Copper-base castings: Brass water fittings are commonly made with special hotbox resin cores.

Health hazards: Formaldehyde is liberated during curing, so coreblowers and the area around them must be well ventilated and exhausted. Avoid skin contact with liquid resins and mixed sand.

Reclamation: It is not normal to reclaim hot box cores when they are used in green sand moulds. Core sand residues in green sand are well tolerated.

General: The hot box process was the first high speed resin coremaking process. Gas-cured cold box processes have superseded it in many cases, particularly for thick section cores, but hot box is still used because of its good surface hardness, strength in thin sections and excellent breakdown properties.

Warm box process

Principle: The binder is a reactive, high furfuryl alcohol binder. The catalyst is usually a copper salt of sulphonic acid. Sand, binder and catalyst are mixed and blown into a heated core box. The heat activates the catalyst which causes the binder to cure.

Sand: Clean silica sand of AFS 50–60 is used. Low acid demand is advisable.

Resin addition: 1.3–1.5% resin and 20% catalyst (based on resin) depending on sand quality.

Nitrogen content: The resin contains about 2.5% N. The catalyst is N-free.

Mixing procedure: Continuous or batch mixers. Catalyst should be added first, then resin.

Bench life: Typically 8 hours.

Core blowing: The binder has low viscosity and blow pressures of around 500 kPa (80 psi) may be used.

Core boxes: Cast iron or steel heated to 180–200°C. Use a silicone based parting agent. A release agent is sometimes added to the sand/binder mix.

Curing time: 10–30 seconds depending on the thickness. Thick section cores continue to cure after ejection from the core box.

Core strength: Surface hardness and strength is high on ejection. Final tensile strength can be 3000–4000 kPa (400–600 psi).

Casting characteristics: Gas evolution is low, the low nitrogen content reduces incidence of gas related defects in ferrous castings. Surface finish of castings is good with low incidence of veining defects. Breakdown after casting is good.

Environmental: Emission of formaldehyde is low at the mixing and curing stages. Avoid skin contact with binder, catalyst and mixed sand.

Reclamation: Core residues in green sand are not harmful.

General: Warm box cores have high strength and resistance to veining and are used in critical applications, usually for iron castings such as ventilated brake discs.

Oil sand

Principle: Certain natural oils, such as linseed oil, known as 'drying oils', polymerise and harden when exposed to air and heat. Natural oils can be chemically modified to accelerate their hardening properties. Silica sand is mixed with the drying oil, a cereal binder and water. The resulting mixture is either manually packed or blown into a cold core box. The cereal binder, or sometimes dextrin, gives some green strength to the core which is then placed in a shaped tray or 'drier' to support it during baking. The baking hardens the oil and the core becomes rigid and handleable.

Sand: Clean silica sand, AFS 50-60.

Binder addition: 1–2% of drying oil
1–2% pre-gelatinised starch $\left.\right\}$ or 2% dextrin
2–2.5% water   0.5–1% water

Nitrogen content: Zero.

Mixing: Batch mixers are preferred in order to develop the green bond strength, mixing times may be 3–10 minutes.

Bench life: As long as the mixture does not lose moisture, the bench life is up to 12 hours.

Core blowing: Oil sand mixtures are sticky and difficult to blow, the highest blow pressure possible (around 700 kPa, 100 psi) is used. They are frequently hand-rammed into core boxes.

Core boxes: May be wood, plastic or metal. Wooden boxes require a paint or varnish to improve the strip. Metal boxes are preferably made of brass or bronze to aid stripping of the fragile green cores.

Curing: A recirculating air oven is needed since oxygen is necessary to harden the oil. The temperature is normally 230°C, allowing 1 hour for each 25 mm section thickness. Burning and consequent friable edges may occur on thin section cores, if this happens, a lower temperature for a longer time should be used.

Core strength: Green strength is low so sagging will occur if the cores are not supported during baking. Correctly baked cores develop tensile strength of 1340 kPa (200 psi).

Casting characteristics: Breakdown after casting is excellent. The gas evolution is high, particularly if cores are under-baked, venting is often necessary. Water based coatings can be used. While for most applications, oil sand has been superseded by synthetic resin processes, it still remains a valuable process in applications where particularly good breakdown of cores is needed, for example for high conductivity copper and high silicon iron castings in which hot tearing is a serious problem.

Health hazards: Unpleasant fumes are emitted during baking and ovens must be ventilated. Some proprietary core oils contain a proportion of mineral oil, which may be harmful if skin contact is prolonged.

Gas triggered processes

Phenolic-urethane-amine gassed (cold box) process

Foseco products: POLITEC and ESHAMINE cold box resin

Principle: The binder is supplied in two parts. Part 1 is a solvent based phenolic resin, Part 2 is a polyisocyanate, MDI (methylene di-phenyl di-isocyanate) is a solvent. The resins are mixed with sand and the mixture blown into a core box. An amine gas (TEA, triethylamine or DMEA, dimethyl ethyl amine or DMIA, dimethyl isopropyl amine) is blown into the core, catalysing the reaction between Part 1 and Part 2 causing almost instant hardening.

Sand: Clean silica sand of AFS 50–60 is usually used but zircon and chromite sands can be used. The sand must be dry, more than 0.1% moisture reduces the bench life of the mixed sand. High pH (high acid demand) also reduces bench life. Sand temperature is ideally about 25°C, low temperature causes amine condensation and irregular cure. High temperature causes solvent loss from the binder, and loss of strength.

Resin addition: Total addition is 0.8–1.5% depending on sand quality. Normally equal proportions of Part 1 and Part 2 are used.

Nitrogen content: Part 2, the isocyanate, contains 11.2% N.

Mixing procedure: Batch or continuous mixers can be used, add Part 1 first then Part 2. Do not overmix since the sand may heat and lose solvents.

Bench life: 1–2 hours if the sand is dry.

Core blowing: Use low pressure, 200–300 kPa (30–50 psi), the blowing air must be dry, use a desiccant drier to reduce water to 50 ppm. Sticking of sand and resin to the box walls can be a problem due to resin blow-off. The lowest possible blow pressure should be used. Use of a special release agent such as STRIPCOTE is advised.

Core boxes: Iron, aluminium, urethane or epoxy resin can be used. Wood is possible for short runs. Use the minimum vents that will allow good filling, since reduced venting gives better catalyst gas distribution. The exhaust vent area should be 70% of the input area to ensure saturation of the core. Core boxes must be sealed to allow the amine catalyst gas to be collected.

Gas generators: The amine catalysts are volatile, highly flammable liquids. Special generators are needed to vaporise the amine and entrain it in air or CO_2. The carrier gas should be heated to 30–40°C to ensure vaporisation. Controlled delivery of amine by pump or timer is desirable.

Gas usage: Approximately 1 ml of amine (liquid) is needed per kilogram of sand. The amine usage should be less than 10% of Part 1 resin. DMEA is faster curing than TEA. Typical curing is a short gassing time (1–2 seconds) followed by a longer (10–20 seconds) air purge to clear residual amine from the core. Over-gassing is not possible but simply wastes amine.

Typical gassing times:

core wt (kg)	total gas+purge (s)
10	10
50	30
150	40

Core strength: Tensile strength immediately after curing is high, 2000 kPa (300 psi), transverse strength 2700 kPa (400 psi). Storage of cores at high humidity reduces the strength considerably. The use of water based coatings can cause loss of strength.

Casting characteristics

Ferrous castings

Good surface and strip without coatings
Low hot strength but this can be improved by adding iron oxide
Tendency to finning or veining
Lustrous carbon formation may cause laps and elephant skin defects on upper surfaces of castings. Addition of red iron oxide (0.25–2.0%) or a coarse-grained form of iron oxide at 1.0–4.0% reduces defects
Breakdown is good
The N content may cause problems on some steel castings.

Aluminium castings

Good surface and strip
Poor breakdown, the resin hardens when heated at low temperature. Aluminium castings may need heat treatment at 500°C to remove cores
There are no hydrogen problems.

Reclamation: Excessive contamination of green sand with cold box core residues can cause problems. Cold box cores and moulds can be thermally reclaimed if uncontaminated with iron oxide.

Environmental: TEA, DMIA and DMEA have objectionable smells even at 3 or 6 ppm. Good, well sealed core boxes, good exhaust and good exhaust gas scrubbers are necessary. The cores must be well purged or amine will be released on storage. Liquid amine is highly flammable, treat like petrol. Air/amine mixtures may be explosive. MDI (Part 2) acts as a respiratory irritant and may cause asthmatic symptoms but it has low volatility at ambient temperature, and is not normally a problem. Avoid skin contact with resins or mixed sand.

General: In spite of its environmental and other problems, the cold box process is so fast and produces such strong cores that it is the most widely used gas triggered process for high volume core production. Good engineering has enabled the environmental problems to be overcome. Reduced free-phenol resins are being produced to assist with sand disposal problems.

The ESHAMINE Plus process

This patented process is a development of the amine gassed phenolic-urethane cold box process which overcomes the problems of using volatile liquid amine catalysts. The liquid amines TEA, DMEA and DMIA must be vaporised to function effectively. Under cold ambient conditions, amine condensation in the gas supply lines or in the core box itself may occur.

In the ESHAMINE Plus process, a gaseous amine, TMA (trimethylamine) is used as catalyst. TMA is the most reactive of the tertiary amine family and has a boiling point of 3°C (compared to 89°C for TEA, 65°C for DMIA and 35°C for DMEA). Being a gas at ambient temperature simplifies the design of gassing equipment, reduces gas consumption and improves productivity. Gassing times are reduced by up to 78% and purge times by up to 54% compared to DMEA, DMIA or TEA catalysed cores. These reductions have been found to have a significant impact on the overall cycle time of the core machine, raising core output by 30%. A further benefit of the reduced gas usage is the lower cost of amine scrubber maintenance.

ECOLOTEC process (alkaline phenolic resin gassed with CO_2)

Foseco product: ECOLOTEC resin

Principle: ECOLOTEC resin is an alkaline phenol-formaldehyde resin containing a coupling agent. The resin is mixed with sand and the mixture blown into a core box. CO_2 gas is passed through the mixture, lowering the pH and activating the coupling agent which causes crosslinking and hardening of the resin. Strength continues to develop after the core is ejected as further crosslinking occurs and moisture dries out.

Sand: Clean silica sand of AFS 50-60 is used. The sand should be neutral (pH

7) with low acid demand. Zircon and chromite sands can be used, but olivine sand is not suitable. Temperature is ideally 15–30°C.

Resin addition: 2.0–2.5% depending on the sand grade.

Nitrogen content: Zero.

Mixing procedure: Batch or continuous.

Bench life: Curing occurs only by reaction with CO_2. The CO_2 in the air will cause a hardened crust on the surface of the mixed sand, but the soft sand underneath can be used. Keep the mixed sand covered. The bench life is usually 5 hours at 15°C reducing to 1 hour at 30°C.

Core blowing: Blow pressure 400–550 kPa (60–80 psi) is needed.

Core boxes: Wood, metal or plastic. Clean them once per shift and use SEPAROL or a silicone based release agent.

Gassing: Cores are blown at 400–550 kPa (60–80 psi) then gassed for 20–40 seconds at 100–300 litres/minute. CO_2 consumption is about 2% (based on weight of sand). Gas should not be forced through the core box at high velocity since the gas must react with the binder. No purge is necessary before extracting the core.

Core strength

	as gassed (kPa)	(kgf/cm^2)	(psi)
Transverse	800–1000	8–10	110–140

Transverse strength doubles after one hour. Cores should be stored in dry conditions.

Casting characteristics: Steel, iron, copper based and light alloy castings can be made. ECOLOTEC is free from N, S and P. Finning and lustrous carbon defects do not occur. Breakdown after casting is good.

Coating: Water or solvent based coatings can be used without affecting the strength. Methanol coatings should be avoided.

Environmental: Gassed cores have no odour. Fume is low at mixing, casting and knockout.

Reclamation: Not normally practised when cores are used in green sand moulds. The inflow of alkaline salts into green sand can result in strong activation of the bentonite clay, a move to a partially activated bentonite will counteract the effect.

General: Tensile strength is not as high as the amine gassed isocyanate process but the excellent casting properties, freedom from nitrogen, lustrous carbon and finning defects and above all its environmental friendliness make ECOLOTEC an attractive process particularly for larger cores and moulds for iron steel and aluminium castings.

The SO_2 process

Principle: Sand is mixed with a furane polymer resin and an organic peroxide, the mixture is blown into the core box and hardened by passing sulphur dioxide gas through the compacted sand. The SO_2 reacts with the peroxide forming SO_3 and then H_2SO_4 which hardens the resin binder.

Sand: Clean silica sand of AFS 50-60. Other sands may be used if the acid demand value is low. The temperature should be around 25°C, low temperature slows the hardening reaction.

Resin addition: Typically 1.2–1.4% resin, 25–60% (based on resin) of MEKP (methyl ethyl ketone peroxide).

Nitrogen content: Zero.

Mixing: Batch or continuous mixers, add resin first then peroxide. Do not overmix since the sand may heat and reduce bench life.

Bench life: Up to 24 hours.

Core blowing: Blowing pressure 500–700 kPa (80–100 psi).

Core boxes: Cast iron, aluminium, plastics or wood can be used. A build up of resin film occurs on the core box after prolonged use, so regular cleaning is needed by blasting with glass beads or cleaning with dilute acetic acid.

Curing: The SO_2 gas is generated from a cylinder of liquid SO_2 fitted with a heated vaporiser. The gas generator must also be fitted with an air purge system so that the core can be cleared of SO_2 before ejection. SO_2 is highly corrosive and pipework must be stainless steel, PTFE or nylon. For large cores, a separate gassing chamber may be used in which the chamber pressure is first reduced then SO_2 is injected and finally the chamber is purged.

Gas usage: 2 g (ml) liquid SO_2 is needed per kg of sand. Cores are normally gassed for 1–2 seconds followed by 10–15 seconds air purge. Overgassing is not possible.

Core strength: Tensile strength is 1250 kPa (180 psi) after 6 hours. Storage properties are good.

Casting characteristics: Coatings are not normally necessary and surface quality of castings is good. Breakdown of cores is excellent with ferrous castings. It is also good with aluminium castings, better than cold box cores, and this has proved to be one of the best applications. The sulphur catalyst may cause some metallurgical problems on the surface of ductile iron castings.

Reclamation: Thermal reclamation is possible. Mechanically reclaimed sand should not be used for SO_2 core making but may be used with furane self-setting sand.

Environmental: SO_2 has an objectionable smell and is an irritant gas. Core boxes must be sealed and exhaust gases must be collected and scrubbed with sodium hydroxide solution. Cores must be well purged to avoid gas release during storage.

MEKP is a strongly oxidising liquid and may ignite on contact with organic materials. Observe manufacturer's recommendations carefully. Avoid skin contact with resin and mixed sand.

SO$_2$ cured epoxy resin

Principle: Modified epoxy/acrylic resins are mixed with an organic peroxide, the mixture is blown into the core box and hardened by passing sulphur dioxide gas through the compacted sand. The SO_2 reacts with the peroxide forming SO_3 and then H_2SO_4 which hardens the resin binder.

Sand: Clean silica sand of AFS 50-60. Other sands may be used if the acid demand value is low. The temperature should be around 25°C, low temperature slows the hardening reaction.

Resin addition: Typically 1.2–1.4% resin, 25–60% (based on resin) of MEKP (methyl ethyl ketone peroxide).

Nitrogen content: Zero.

Other details: Similar to SO_2/furane process. Bench life is up to 24 hours.

Ester-cured alkaline phenolic system

Foseco product: FENOTEC resin

Principle: The resin is an alkaline phenolic resin (essentially the same as the self-hardening resins of this type). Sand is mixed with the resin and blown or manually packed into a corebox. A vaporised ester, methyl formate, is passed through the sand, hardening the binder.

Sand: Highest strengths are achieved with a clean, high silica sand of AFS 50-60. Sand temperature should be between 15 and 30°C.

Resin addition: Typically 1.5% total.

Nitrogen content: Zero, sulphur is also zero.

Mixing procedure: Batch or continuous.

Bench life: Curing only occurs by reaction with the ester-hardener, so the bench life is long, 2–4 hours.

Core blowing: Blow pressure 350–500 kPa (50–80 psi) is needed.

Core boxes: Wood, plastic or metal. Clean once per shift and use SEPAROL or a silicone based release agent.

Gassing: Methyl formate is a colourless, highly flammable liquid, boiling at 32°C. It has low odour. A specially designed vaporiser must be used to generate the methyl formate vapour. Usage of methyl formate is about 20–30% of the resin weight. Core boxes and gassing heads should be sealed correctly and the venting of the core box designed to give a slight back pressure so that the curing vapour is held for long enough for the reaction to take place.

Core strength: Compression strengths of 5000 kPa (700 psi) are possible.
    Tensile and transverse strengths are not available, they are not as high as phenolic isocyanate resins.

Casting characteristics: Finning and lustrous carbon defects are absent. Good high temperature erosion resistance and good breakdown. Mostly used for steel and iron castings.

Environmental: Low odour, but methyl formate is flammable and care is needed.

Reclamation: Reclamation by attrition is possible but the reclaimed sand is best re-used as a self-hardening alkaline-phenolic sand where strengths are not as critical. Core sand residues entering green sand causes no problems.

General: Strengths are relatively low so the process has mainly been used for rather thick section moulds and cores where high handling strength is not necessary.

Review of resin coremaking processes

The heat activated resin coremaking processes, Croning/Shell and hot box were developed in the 1950s and 1960s and rapidly supplanted the older oil sand coremaking process. The attraction was speed of cure and the fact that cores could be cured in the box, eliminating the dimensional problems of oil sand cores. By the mid-1970s the amine-urethane cold box process was becoming firmly established, bringing great benefits of speed as well as an environmental burden. Since 1976 there have been many further developments in the amine-urethane systems and in other cold gas hardened core binder systems.

Reasons for this intense interest include:

The flexibility of gas-hardened processes which makes them suitable for mass production of cores with automatic equipment

Fast curing of cores in the box by controlled injection of reactive gases

High strengths on ejection, often 80% of the final strength

Dimensional accuracy resulting from cold curing

Availability of an extensive range of specially designed core-making equipment

Energy saving through cold operation

Rapid tooling changes made easy by low tooling temperatures

By 1990 the majority of cores were produced by the cold box, amine isocyanate process (Table 13.1).

Table 13.1   Coremaking processes used in 48 automobile foundries in Germany, 1991

Amine cold box	44
Hot box	10
Shell/Croning	9
CO_2 – silicate	3

Tables 13.2a and b summarise the production features of the many coremaking processes in use.

Figure 13.5 shows how the use of chemical binders has changed in the USA over the last 40 years.

Shell sand is declining slightly, holding up remarkably for such an old process.

Core oil has declined in the face of hot box and cold box.

Silicates received a boost in 1980, due to the introduction of ester-hardening, but have declined slowly since.

Hot box has declined due to the rise in use of cold box phenolic urethane core binders, but still has a significant place.

Acid cured furanes have held steady since the maximum in 1980.

Urethane no-bakes are popular in the US, although their use in Europe is limited.

New binders are predicted to increase substantially by the year 2000.

Table 13.2a Production features of core binders

Process	Core-box requirements	Curing	Typical binder (%)	Flowability	Breakdown	Integration into green sand
Shell	Metal, heated	260°	2–5	1	2	2
Hot box	Metal, heated	230°C	1.2–2.5	2	1	2
Warm-box	Metal, heated	150–180°C	1.2–2.0	2	1	2
Amine-urethane	Any material, vented, sealed	Amine gas	0.9–2.0	1	1	2
SO$_2$-epoxy	Any material, vented, sealed	SO$_2$ gas and peroxide	1.0–1.2	1	1	2
SO$_2$-furane	Any material, vented, sealed	SO$_2$ gas	1.0–1.5	1	1	2
Methyl-formate Alkaline phenolic	Any material, vented, sealed	Methyl formate gas	1.4–1.6	1	2	1
CO$_2$-silicate	Any, vented	CO$_2$ gas	2.5–3.5	3	3	1
ECOLOTEC	Any, vented	CO$_2$ gas	2.0–2.5	1	2	1

1 Good – unlikely to restrict application
2 Satisfactory – suitable for most applications
3 Restrictions – may limit applications

Table 13.2b Properties of core binders

Process	Strength	Gas evolution properties	Typical bench life	Typical strip time	Minimum strip to pour time	Pin-hole tendency	Veining	Lustrous carbon
Shell	1	3	Indefinite	2 mins	30 min.	3	1	1
Hot box	1	2	4–6 hrs	1/2–1 min.	1/2–2hrs	3	2	1
Warm box	1	1	4–8 hrs	1/2–1 min.	0–2 hrs	3	1	1
Amine-urethane	1	1	1–6 hrs	10–30 sec.	0–1 hr	2	2	2
SO_2-epoxy	1	2	24 hrs	10–30 sec.	Instant	1	1	1
SO_2-furane	2	2	24 hrs	10–30 sec.	Instant	2	2	2
CO_2-silicate	3	1	2–3 hrs	10–60 sec.	1 hr	1	1	1
ECOLOTEC	2	1	3–4 hrs	5–30 sec.	Instant	1	1	1

1 Good – unlikely to restrict application
2 Satisfactory – suitable for most applications
3 Restrictions – may limit applications

U.S. foundry industry consumption of sand binder by system

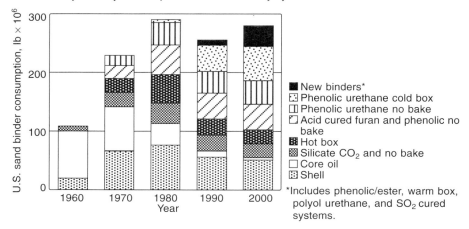

■ New binders*
⊡ Phenolic urethane cold box
⊞ Phenolic urethane no bake
▱ Acid cured furan and phenolic no bake
▦ Hot box
▨ Silicate CO_2 and no bake
☐ Core oil
⊟ Shell

*Includes phenolic/ester, warm box, polyol urethane, and SO_2 cured systems.

Figure 13.5 *US foundry industry consumption of sand binder by system. (From* Foundry Management & Technology, *Jan. 1995 p. D-5).*

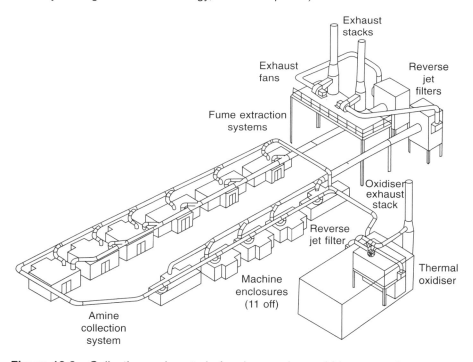

Figure 13.6 *Collection and control of amine gas in a cold box core shop.*

The environmental difficulties of the amine-isocyanate process have been solved, at a cost, by engineering. Figure 13.6 shows how a large coreshop collects and disposes of amine gas. Smaller foundries find the problems of handling amine gas difficult so there is a need for a more environmentally friendly process. Up to the present, only the CO_2-hardened ECOLOTEC system meets the requirements.

A growing concern of all users of chemical binders is that of disposing of used sand. All the developed countries impose severe restrictions on the materials that can be safely dumped on landfill sites. Pollution of water supplies by phenol and other leachable chemicals is a growing concern, so all resin manufacturers are reducing the free phenol content of their products. The use of sand reclamation, both mechanical and thermal is being increasingly used by foundries to reduce the quantity of sand that must be disposed to waste sites.

Chapter 14
Sodium silicate bonded sand

Sodium silicate is a water soluble glass available from suppliers in a wide range of types specified by the silica (SiO_2), soda (Na_2O) and water content. Manufacturer's data sheets specify the 'weight ratio' of silica to soda, the percentage of water and the viscosity. For foundry use, sodium silicates with ratios between 2 and 3 and water content around 56% are usually used.

Sodium silicate can be hardened in a number of ways: by adding weak acids (CO_2 gas or organic esters), by adding various powders (di-calcium silicate, anhydrite etc.) or by removing water. CO_2 gas and liquid ester hardeners are the most widely used of the silicate processes. Silicate binders have no smell and few health hazards, but the bond strength is not as high as that of resin binders and, being an inorganic bond, it does not burn out with heat, so breakdown after casting can be a problem. For this reason, various organic breakdown additives are often incorporated with the liquid silicate or added during sand mixing.

Typical data for foundry grade sodium silicate:

	Wt ratio	Na_2O (%)	SiO_2 (%)	H_2O (%)	s.g.	Visc.cP at 20°C	litres/ tonne
Low ratio	2.0	15.2	30.4	54.4	1.56	850	641
Medium ratio	2.40	12.7	30.8	56.5	1.50	310	668
High ratio	2.85	11.2	31.9	56.9	1.48	500	677

CO_2 silicate process (basic process)

Principle: Sand is mixed with sodium silicate and the mixture blown or hand-rammed into a core box or around a pattern. Carbon dioxide gas is passed through the compacted sand to harden the binder.

Sand: Silica sand of AFS 50-60 is usually used, the process is quite tolerant of impurities and alkaline sand such as olivine can be used. Sand should be clay free. Low temperature retards hardening, the sand temperature should be above 15°C.

Binder addition: About 3.0–3.5% silicate addition is usually used. High ratio silicates give short gassing times, low gas consumption and improved post-casting breakdown, but high ratio silicate is easily over-gassed, resulting in reduced core strength and poor core storage properties. Low ratio binders give better core storage, but gassing times are longer and breakdown is not as good. Low ratio silicates (2.0–2.2) are usually used.

Mixing: Batch or continuous mixers are used, avoid excessive mixing time which may heat the sand and cause loss of water.

Bench life: Several hours if drying is prevented by storing in a closed container or covering with a damp cloth.

Core boxes: Any material; wood, resin or metal may be used but the boxes must be painted to prevent sticking. Polyurethane or alkyd paints are used followed by the application of a wax or silicone-wax polish, or STRIPCOTE AL.

Blowing: High pressure is needed, 650–700 kPa (90–100 psi).

Curing: The correct volume of gas is needed. Undergassing does not achieve optimum strength. Overgassing leads to reduction of strength on standing. To ensure correct hardening, a flow controller and timer should be used. If cores are to be stored for some time before use, it is preferable to undergas. Higher strength will then be developed on standing. Gassing times are typically 10–60 seconds, depending on the size of the core.

CO_2 gas consumption per kg of sodium silicate (good quality silica sand AFS 50)

Silicate ratio	Time to reach 700 kPa (100 psi) compression strength at 5 l/min	litres CO_2	kg CO_2
2.0	52 s	300	0.60
2.2	48	275	0.55
2.4	26	150	0.30
2.5	23	130	0.26

Core strength:

	as-gassed		after 6 h	
Tensile	350 kPa	50 psi	1000 kPa	150 psi
Compressive	2100	300	4200	600

Core storage: Overgassing causes cores to become friable during storage over 24 hours. Careful control of gassing is essential (Fig. 14.1).

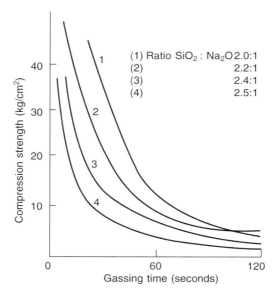

Figure 14.1   *The effect of 24 hour storage of cores gassed for different times as a function of silicate ratio.*

Casting characteristics: Cores require coating to produce good surface finish on ferrous castings. Spirit based coatings should be used because water can soften the silicate bond. No metallurgical problems occur with ferrous or non-ferrous castings, the binder contains no nitrogen or sulphur. Breakdown after casting is poor (with ferrous castings) because silicate fuses with sand above 400°C.

Reclamation: Reclamation is difficult because the binder does not burn out during casting.

Health hazards: There is very little fume or smell during core manufacture, storage or casting. Gloves should be worn when handling silicates or mixed sand.

General comments: The CO_2 process was the first of the gas-hardened processes to be introduced into the foundry and is still one of the easiest and cleanest processes to use. Foseco and other suppliers have made important improvements to the process while still retaining its ease of use and excellent environmental characteristics.

Gassing CO_2 cores and moulds

The core boxes are prepared by coating with a suitable parting agent, silicone waxes can be used but the best results are obtained with STRIPCOTE AL, an aluminised self-drying liquid.

The simplest gassing techniques are based either on a simple tube probe fitted through a backing plate, or a hood sealed to the open side of the box with a rubber strip (Fig. 14.2). Gas at a pressure of about 70–140 kPa (10–20 psi) is passed until the sand is felt to have hardened or some white crystals appear (indicating overgassing). Venting of the boxes and patterns to prevent dead spots is essential. It is usual to rap the box before gassing to make stripping easier.

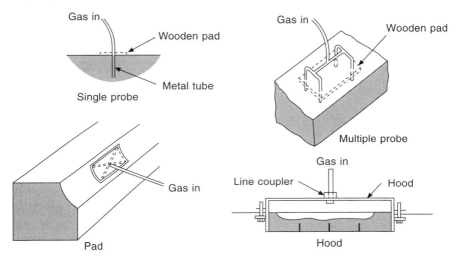

Figure 14.2 *CO$_2$-silicate process, methods of gas application.*

Control of gassing is improved by incorporating gas heaters (to prevent freezing of the valves), flow meters and timers so that automatic control of the volume of gas passed is possible. Not only does this practice economise on expensive CO$_2$, but the best strength and core-storage life will only be achieved if the correct amount of gas is used. It is important that overgassing is avoided, any error should always be towards undergassing. The suppliers of CO$_2$ gas will advise on the best equipment to use.

Improvements to the CO$_2$ silicate process

Foseco products: CARSIL } sodium silicate blended with special
SOLOSIL } additions
DEXIL   breakdown agent

Principle: The main drawbacks of the basic CO$_2$ process are:

poor breakdown of the bond after casting
poor core storage properties
rather low tensile strength

These properties can be greatly improved by special additives, while retaining the simplicity and user friendliness of the CO_2 process.

The CARSIL and SOLOSIL range: These products are a range of sodium silicate based binders for the CO_2 process. They may be simple sodium silicates which can be used with DEXIL breakdown agent if required, or they may be 'one-shot' products which incorporate a breakdown agent, or they may (like SOLOSIL) incorporate special additives to improve bond strength as well as breakdown.

Binders containing high levels of breakdown additives give improved post-casting breakdown but the maximum as-gassed strength is reduced and core storage properties are likely to be impaired. The selection of an optimum binder for a given application is therefore almost always a compromise. The requirement for high production rates and high as-gassed strength must be balanced against core storage properties and the need for good breakdown. The range of binders includes some which are suitable only for the CO_2 process, some which are suitable for self-setting applications and some which can be used for both processes.

The commonly used breakdown agents are organic materials which burn out under the effect of the heat of the casting. While solid breakdown agents such as dextrose monohydrate, wood flour, coal dust and graphite can be used, powder materials are not easy to add consistently to sand in a continuous mixer. Liquid breakdown agents are easier to handle, they usually consist of soluble carbohydrates. The best improve gassing speed without loss of strength. Some are also resistant to moisture pick-up and their use has increased the storage life of high-ratio silicate bonded cores.

Sucrose is the only common carbohydrate soluble in sodium silicate without a chemical reaction. It is readily soluble up to 25% and many sugar or molasses-based binders are available. Use of sucrose increases gassing speed but reduces maximum strength and storage properties. Nevertheless silicates containing sugar are the most popular CO_2 binders because of the convenience of a binder in the form of a single liquid. Molasses can be used as a low cost alternative to sugar, but it is subject to fermentation on storage. The Foseco CARSIL range of silicate binders is based on sugar. Some are designed for use with CO_2, others for self-setting (SS) with ester hardeners. Some can be used for both processes.

The CARSIL range of silicate binders

The extent to which a core will break down after casting varies depending on the type of metal cast. Low temperature alloys such as aluminium do not inject enough heat into the sand to burn out the breakdown agent fully. Indeed, the low temperature heating may even strengthen the core. In such cases it is useful to add additional breakdown agents such as DEXIL.

Product	Ratio	Additive	CO$_2$/SS	Comments
CARSIL 100	2.5:1	Sugar	CO$_2$/SS	Higher ratio for faster gassing, take care not to overgas. Can be used with ester hardeners.
CARSIL 513	2.4:1	Sugar	CO$_2$/SS	Low viscosity binder for easy mixing in continuous mixers. Moulds and cores.
CARSIL 520	2.0:1	High sugar	CO$_2$	High breakdown, low viscosity.
CARSIL 540	2.2:1	Low sugar	CO$_2$	Suitable for moulds or cores.
CARSIL 567	2.2:1	High sugar	CO$_2$	High breakdown, good for AI casting.

Note: Some of the CARSIL binders were formerly known as GASBINDA binders in the UK.

DEXIL 34BNF is a powder additive developed for use with light alloys. It also acts as a binder extender so reducing the silicate requirement. The application rate is 0.5–1.5%. It should be added to the sand and pre-dispersed before adding the silicate.

DEXIL 60 is a pumpable organic liquid which is particularly suitable for use with continuous mixers.

SOLOSIL

SOLOSIL was developed to improve on the performance of silicates containing sugar based additives. SOLOSIL is a complex one-shot sodium silicate binder for the CO$_2$ gassed process. It contains a high level of breakdown agent/co-binder and offers a combination of high strength and rapid gassing with good core storage properties and excellent post-casting breakdown.

The binder is best used with good quality silica sand. Addition levels of 3.0–4.5% are used depending on the application. To take full advantage of the high reactivity, an automatic gassing system incorporating a vaporiser, pressure regulator, flow controller and gassing timer is advisable. The high rate of strength development is shown in Fig. 14.3. While the transverse and tensile strength developed by SOLOSIL binders are still somewhat lower than some organic resin binders, SOLOSIL generally proves more cost effective and overcomes problems of poor hot strength, veining and finning, gas pinholing and fume on casting which occur with some resin binders.

Self-setting sodium silicate processes

The first self-setting process used powder hardeners. The Nishiyama process used finely ground ferrosilicon powder which reacts with sodium silicate generating heat and forming a very strong bond. The reaction also generates hydrogen which is dangerous. Other powder hardeners (which do not evolve dangerous gases) include di-calcium silicate, certain cements (such as blast

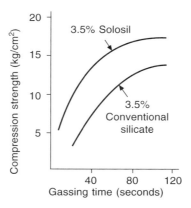

Figure 14.3   *Strength development of SOLOSIL compared with a conventional sugar/silicate system*

furnace cement and sulphate resisting cement) and anhydrite. However, all powder hardeners are difficult to add uniformly to sand in continuous mixers, and their reactivity is difficult to control, since particle size and the age after grinding affect the reactivity of the powder. When liquid hardeners based on organic esters were introduced, the use of powder hardeners was largely discontinued.

Ester silicate process

Foseco products: CARSIL   sodium silicate binders
CARSET   ester hardener
VELOSET   special ester for very rapid setting

Principle: Sand is mixed with a suitable grade of sodium silicate, often incorporating a breakdown agent, together with 10–12% (based on silicate) of liquid organic ester hardener. The acid ester reacts with and gels the sodium silicate, hardening the sand. The speed of hardening is controlled by the type of ester used.

Sand: Dry silica sand of AFS 45-60 is usually used. As with all silicate processes, the quality and purity of the sand is not critical, alkaline sand such as olivine can be used. Fines should be at a low level. Sand temperature should be above 15°C, low temperature slows the hardening.

Additions: Sodium silicates with ratios between 2.2 and 2.8 are suitable, the higher the ratio, the faster the set. Silicates containing breakdown agents are usually used, additions between 2.5 and 3.5% are used depending on the sand grade. The ester hardener is commonly:

glycerol diacetate   fast cure
ethylene glycol diacetate   medium cure
glycerol triacetate   slow cure

Proprietory hardeners may be blends of the above with other esters. The addition level is 10–12% of the silicate.

Pattern equipment: Wood, resin or metal patterns can be used. Core boxes and patterns should be coated with polyurethane or alkyd paint followed by application of wax polish. STRIPCOTE parting agent may also be used.

Mixing: Continuous mixers are usually used, if batch mixers are used, the ester hardener should be mixed with the sand before adding the silicate.

Speed of strip: 20–120 minutes is common with normal ester hardeners. Attempts to achieve faster setting may result in lower strength moulds because the work time becomes short. With certain esters there is a tendency for core and mould distortion due to sagging if stripping occurs too early. Faster setting can be achieved by using the special VELOSET hardener.

Strength: The final strength achieved is:

Tensile   700 kPa (100 psi)
Compression   2000–5000 kPa (300–700 psi)

Coatings: Spirit based coatings should be used.

Casting characteristics: No metallurgical problems arise with ferrous or non-ferrous castings. Breakdown is poor unless a silicate incorporating a breakdown agent is used.

Reclamation: As with all silicate processes, burnout of the bond does not occur during casting and attrition does not remove all the silicate residue so that build-up occurs in the reclaimed sand, reducing refractoriness and leading to loss of control of work time and hardening speed. The VELOSET system has been specially developed to permit reclamation (see below).

Environment: Silicate and ester have little smell and evolve little fume on casting. Silicates are caustic so skin and eye protection is needed while handling mixed sand.

CARSET 500 Hardeners: These are blends of organic esters formulated to give a wide range of setting speeds when used with sodium silicates, particularly the CARSIL series of silicates which incorporate a breakdown agent. For the best results, the silicate addition should be kept as low as possible in relation to the sand quality and the CARSET hardener maintained at 10% by weight of the silicate level. The speed of set is dependent on the sand temperature, silicate ratio and grade of CARSET hardener used.

The CARSET 500 series of hardeners

CARSET 500 series	Gel times (minutes) at 20°C		
	CARSIL 540 2.2 ratio	CARSIL 513 2.4 ratio	CARSIL 100 2.5 ratio
500	8	7	5
511	9	8	6
522	13	12	8
533	19	15	9
544	105	53	21
555	–	–	90

Note: The gel time is the time taken for gelling to occur when silicate liquid is mixed with an appropriate amount of setting agent. The setting times may not be repeated exactly when sand is present, due to the possibility of impurities, but it does provide a useful guide.

VELOSET hardeners: The VELOSET range is a series of advanced ester hardeners for the self-setting silicate process. They have been designed to give very rapid setting speed with a high strength, excellent through-cure and a high resistance to sagging. Used in the VELOSET Sand Reclamation Process, they provide the only ester silicate process in which the sand can be reclaimed by a simple dry attrition process and reused at high levels equal to those typical of resin bonded sands.

Additions: There are three grades of VELOSET hardener. VELOSET 1, 2 and 3 Binders of ratio 2.2–2.6 are used, lower ratios give inferior strength while if higher ratios are used the bench life becomes too short. The bench life obtained is independent of addition level. The level is usually 10–12% based on the binder. If the sand is to be reclaimed, the addition level of 11% should not be exceeded.

Bench life (minutes) at 20°C

CARSIL ratio	VELOSET grade		
	1	2	3
2.2	10	7	4
2.4	7	4	2
2.6	4	2	1

When a choice is possible, always use the highest ratio CARSIL binder and the slowest grade of VELOSET hardener. This provides optimum strength development.

Mixer: Since VELOSET is rapid setting, it is preferable to use a continuous mixer.

VELOSET sand reclamation process: With the conventional ester silicate process, dry attrition reclamation has occasionally been practised but the level of sand reuse is rarely more than 50%, which hardly justifies the capital investment involved. With the VELOSET system, up to 90% reuse of sand is possible using mechanical attrition.

The process stages are:

Crushing the sand to grain size
Drying
Attrition
Classification
Cooling

The reclaimed sand is blended with new sand in the proportion 75 to 25. During the first 10 cycles of reuse, the sand system stabilises and the bench life of the sand increases by a factor of up to 2. Also, mould strength should improve, and it is usually possible to reduce the binder addition level by up to 20% yet still retaining the same strength as achieved using new sand. Once the process has become established, it may become possible to reuse up to 85–90% of the sand (Figs 14.4 and 14.5). See also p. 179.

Adhesives and sealants

It is often necessary to joint cores together to form assemblies, or to glue cores to moulds before closing the mould. A range of CORFIX adhesives is available:

CORFIX grade	Type	Set time	Temp (°C)	Remarks
4	Stove hardening	30	180–220	High viscosity gap filling
8	Air hardening	slow	ambient	For CO_2 and self-set silicate
21	Air hardening	fast	ambient	Any cold core
25	Hot melt	open time 15–120 s	140–180	Core assembly at high rates, shell process

CORSEAL sealants

This is a group of core sealing or mudding compounds for filling out joint lines, cracks and minor blemishes in cores. CORSEAL is available in two forms.

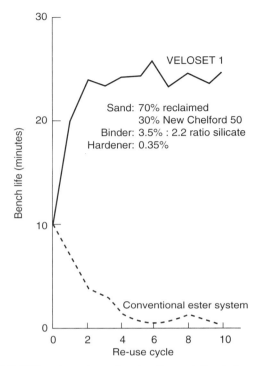

Figure 14.4   *VELOSET reclamation, showing the variation in bench life after repeated use of reclaimed sand, compared with conventional ester process.*

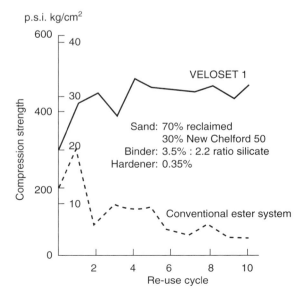

Figure 14.5   *VELOSET reclamation, ultimate strength characteristics of reclaimed sand, compared with conventional ester process.*

CORSEAL 2 is a powder which is mixed with water to form a thick paste (4 parts product to 3 parts water). The paste is applied by spatula or trowel (or fingers) and allowed to dry for about an hour. It may be lightly torched if required immediately.

CORSEAL 3 and 4 are ready-mixed self-drying putties which are sufficiently permeable when fully dry to prevent blowing but strong enough to prevent metal penetration into the joint. Drying time depends on local conditions and the thickness of the layer applied but should be at least 30 minutes.

TAK sealant

Small variations in the mating faces of moulds due to flexing of patterns or deformation of moulding boxes and moulding materials may result in gaps into which liquid metal will penetrate causing runout and flash. This can be prevented by the application of TAK plastic mould sealant which forms a metal and gas-tight seal. TAK does not melt at high temperatures and, if metal touches it, it burns to a compact, fibrous mass. The TAK strip is laid around the upper surface of the drag mould, about 25 mm from the edge of the mould cavity and the mould is then closed and clamped. TAK can also be used to seal small core prints.

TAK 3 is supplied in cartridge form for extrusion from a hand gun, a variety of nozzle sizes is available.

TAK 500 is ready-extruded material supplied in continuous lengths of 6 mm diameter.

Chapter 15

Lost foam casting

Principle of the process

Unlike any other sand casting process, no binders are used. Pre-forms of the parts to be cast are moulded in expandable polystyrene or special expandable copolymers. Complex shapes can be formed by gluing mouldings together. The pre-forms are assembled into a cluster around a sprue then coated with a refractory paint. The cluster is invested in dry sand in a simple moulding box and the sand compacted by vibration. Metal is poured, vaporising the preform and replacing it to form the casting (Fig. 15.1).

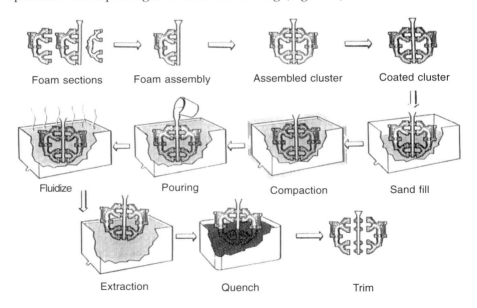

Figure 15.1 *The lost foam casting process.*

Many problems hindered the development of the lost foam casting process. By working closely together, designers, foundries, equipment engineers, polymer manufacturers and coating suppliers have now removed these barriers, making the process a cost effective way to manufacture quality castings. The casting of iron components was initially very difficult due to the formation of lustrous carbon defects on the surface and subsurface carbonaceous inclusions, which were only revealed on machining. Essential to the advancement of ferrous lost foam (LF) casting was the development of copolymers, such as the patented Foseco Low Carbon Bead, which helps to eliminate these defects and so make the process viable.

Patternmaking

The raw bead to be moulded is purchased from a chemical supplier. It consists of spherical beads of polystyrene (EPS) or copolymer of carefully graded size. The bead is impregnated during manufacture with a blowing agent, pentane. The first step is to pre-expand the bead to the required density by steam heating. It is then moulded in a press rather like a plastic injection moulding press. The moulding tool is made of aluminium, hollow-backed to have a wall thickness of around 8 mm. The pre-expanded bead is blown into the closed die, which is then steam heated causing the beads to expand further and fuse together. After fusing, the die is cooled with water sprays (often with vacuum assistance) so that the pattern is cooled sufficiently to be ejected without distortion.

The cycle time is dependent on the heating and cooling of the die, a process which is necessarily rather slow, taking 1–2 minutes for a cycle. The moulding machines are large with pattern plate dimensions typically 800 × 600 mm or 1000 × 700 mm so that multiple impression dies can be used to increase the production rate.

Some lost foam foundries purchase foam patterns from a specialist supplier, such as Foseco–Morval (in the USA and Canada). Others make their own patterns. The casting reproduces in astonishing detail the surface appearance of the foam pattern. A great deal of effort has been put into improving surface quality and special moulding techniques such as Foseco–Morval's 'Ventless Moulding Process' have been developed to minimise bead-trace. EPS mouldings for casting have now reached levels of quality and complexity far beyond that expected from a material designed originally for packaging.

Joining patterns

Where possible, patterns are moulded in one piece using the techniques developed for plastic injection moulding such as metal 'pull-backs' and collapsible cores, but many of the complex shapes needed to make castings cannot be moulded in one piece. Sections of patterns are glued together quickly and precisely using hot melt adhesives using special glue-printing machines (Fig. 15.2). The adhesive is reproduced in the finished casting so that it is important to lay down a controlled amount of glue to avoid an unsightly glue line.

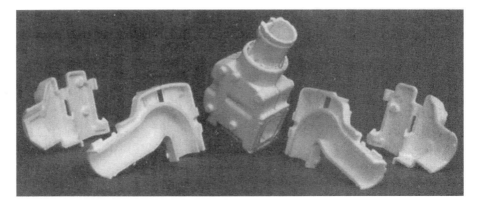

Figure 15.2   *Sections of patterns are glued together to form complex shapes.*

Assembling clusters

Some large castings are made singly, the EPS pattern being attached to a down-sprue of EPS or, in the case of ferrous castings, usually of hollow ceramic fibre refractory. Smaller castings are made in clusters, the patterns being assembled around the sprue with the ingates acting also as supports for the cluster.

Coating the patterns

The foam pattern must be covered with a refractory coating before casting. If no coating is used, sand erosion occurs which can lead to mould collapse during casting. The coatings are applied by dipping and must be thoroughly dried in a low temperature oven before casting.

Investing in sand

The coated pattern clusters are placed in a steel box and dry silica sand poured around. As the box is filled, it is vibrated to compact the sand and cause it to flow into the cavities of the pattern. It was the failure to vibrate correctly that led to so many problems in the early days of lost foam. Following much research by equipment suppliers, good vibration techniques have been developed and patterns with complex internal form can now be reliably invested. It is not possible to persuade sand to flow uphill by vibration so patterns must be oriented in such a way that internal cavities are filled horizontally or downwards.

The mechanism of casting into foam patterns

When low melting point metals such as aluminium alloys are cast into EPS foam, the advancing metal melts and degrades the polystyrene to lower molecular weight polymers and monomers. These residues are then transported through the coating as both liquids and gases into the sand. This process is illustrated in Fig. 15.3. Failure to remove the gaseous residues quickly enough results in slow mould filling and misrun castings, Fig.15.4. If the gases escape too quickly, however, then the metal will fill the cavity in an uncontrolled turbulent manner giving rise to oxide film defects and even mould collapse (Fig. 15.5). If the liquid residues are not absorbed by the coating, a thin carbon film may form on the liquid metal front which, if trapped, may cause a 'carbon-fold' defect (Fig. 15.6).

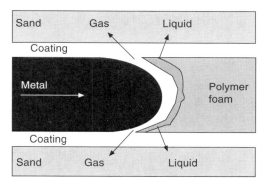

Figure 15.3 *An illustration of the transport of gas and liquid polymer degradation residues through STYROMOL and SEMCO PERM lost foam coatings.*

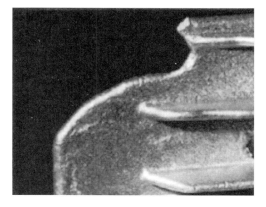

Figure 15.4 *A mis-run defect caused by the use of a coating with insufficient permeability.*

When iron is cast, the higher temperature causes rapid breakdown of the EPS with much gas formation. Vacuum may be applied to the casting flask to aid the removal of the gas. High temperature pyrolysis of EPS results in tar-like carbonaceous products which are not always able to escape from the mould. They break down further to a form of carbon called 'lustrous carbon' which causes a defect characterised by pitting and wrinkling of the

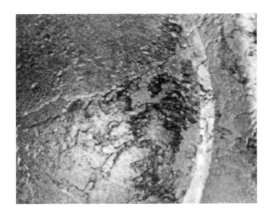

Figure 15.5   *An oxide film defect (on aluminium) caused by turbulent mould filling through use of a coating with too high permeability.*

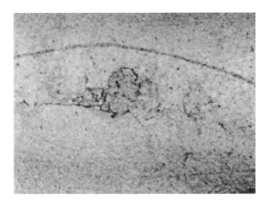

Figure 15.6   *A cold fold defect (on aluminium) caused by poor liquid residue removal.*

upper surface of the casting. High carbon irons, such as ductile iron, are most prone to the defect. In the early 1980s it seemed as though the use of the lost foam process would be seriously restricted by this problem. It was then found that the polymer PMMA (polymethyl methacrylate) depolymerises completely under heat to gaseous monomer leaving no residues in the casting. PMMA is not easy to make in an expandable form and not easy to mould either. Foseco has developed 'Low Carbon Bead', a copolymer of EPS-PMMA which has moulding properties similar to EPS and eliminates lustrous carbon in all but the heaviest section castings (Fig. 15.7).

Foseco and SMC have developed STYROMOL and SEMCO PERM lost foam coatings, designed to remove both the liquid and gaseous residues at the rate required to give controlled mould filling and defect-free castings (see p. 239). These coatings include:

Insulating STYROMOL 169 series coatings for aluminium.
Non insulating STYROMOL and SEMCO PERM coatings for grey and ductile iron.

Figure 15.7 *The use of Foseco Low Carbon Copolymer Bead eliminates lustrous carbon defects on grey and ductile iron castings.*

High refractoriness STYROMOL and SEMCO PERM coatings for chrome irons, manganese and carbon steels.

Methoding

Patterns must be orientated to allow complete filling with sand. Ingates must fulfil the dual role of supporting the fragile pattern cluster and controlling the metal flow. In aluminium casting, it is mainly the permeability of the coating that controls the filling of the casting and very gentle, turbulence free filling is possible with direct pouring. Iron castings, having higher density and heat content, are usually bottom gated to allow controlled mould filling.

Advantages of lost foam casting

1. Low capital cost: The capital cost of a lost foam foundry is as little as 50% of a green sand plant of similar capacity.
2. Low tooling cost: Though tools are expensive, their life is long, so for long-running, high volume parts like aluminium manifolds, cylinder heads and other automotive parts, tool costs are much lower than for gravity or low pressure dies which have a shorter life and require multiple tool sets because of the long cycle time needed for each casting. For shorter running parts the advantage is less and may even be a disadvantage.
3. Reduced grinding and finishing: There is a major advantage on most castings since grinding is restricted only to removing ingates.
4. Reduced machining: On many applications, machining is greatly reduced and in some cases eliminated completely.

5. Ability to make complex castings: For suitable applications, the ability to glue patterns together to make complex parts is a major advantage.
6. Reduced environmental problems: Lost foam is fume free in the foundry and the sand, which contains the EPS residues, is easily reclaimed using a simple thermal process.

Disadvantages

1. The process is difficult to automate completely; cluster assembly and coating involves manual labour unless a complete casting plant is dedicated to one casting type so that specialised mechanical handling can be developed.
2. Methoding the casting is still largely hit and miss and a good deal of experimentation is needed before a good casting is achieved.
3. Cast-to-size can be achieved but only after several tool modifications because the contractions of foam and casting cannot yet be accurately predicted.
4. Because of 2 and 3, long lead times are inevitable for all new castings.

Applications

The successful foundries have been extremely selective in choosing casting applications. In general, lost foam is not regarded as a low-cost method of casting. It is the final cost of the finished component that must be considered.

Aluminium castings

By far the largest proportion of Al castings are used in the automotive industry so this has proved the biggest market for LF.

Inlet manifolds: This was the first successful application and is still being used. The usual process for manifolds is gravity diecasting, a slow process with high tooling cost. Lost foam is used to best advantage in water-jacketted carburettor manifolds.

Cylinder heads: The fastest growing automotive application. Heads are conventionally cast by gravity die with a complex package of cores set into an iron die. Use of LF gives the designer rather more freedom to cool the working face effectively, the combustion chambers can be formed 'as-cast', avoiding an expensive machining operation and bolt holes can be cast.

Cylinder blocks: Automotive companies are moving away from grey iron towards Al blocks. LF offers substantial design advantages; features can be cast in, such as the water pump cavity, alternator bracket, oil filter mounting pad. Oil feed and drain line and coolant lines can also be cast more effectively.

A variety of other automotive parts are being made, water pump housings, brackets, heat exchangers, fuel pumps, brake cylinders. Applications will increase as designers become aware of the design potential. High strength parts such as suspension arms have proved difficult because of metallurgical differences between LF and conventional castings. New developments such as the Castyral Process in which pressure is applied to the solidifying casting are being adopted to improve strength.

Grey iron

Modern automated green sand is such a fast and efficient process that LF cannot compete on cost of the casting alone. Foundries must look for castings where the precision of LF gives savings in machining costs.

Stator cases: This has proved one of the most successful applications. There are weight savings of up to 40% to be made through thinner cooling fins, machining of the bore can be reduced, though not eliminated, mounting feet can be cast complete with bolt holes.

Valves: Grey iron valve bodies, caps and gates are being cast in large numbers. Flange faces are flat so by casting-in bolt holes etc. machining can be eliminated completely in many cases.

Ductile iron

Pipe fittings: One of the most successful applications, the precision of LF is so great that flanges can be cast flat complete with cast bolt holes so that machining can be eliminated altogether. Bends, tees, spigots etc. up to at least 300 mm diameter are being cast in large volume, replacing resin sand moulding. Achieving success has not been easy because of distortion problems both on foam patterns and castings, but the potential benefits are so great that foundries in Europe and Japan have persisted and finally achieved success.

Valves: Machining can be eliminated completely in many cases, particularly with redesign of the part. LF is becoming the accepted way of making water valves up to about 150 mm pipe diameter.

Hubs: Lower weight and reduced machining is possible.

Differential cases: Several automotive companies have persisted with this difficult casting. The advantages are reduced machining (on the internal surfaces) and better balance.

Manifolds: Weight savings are achieved through control of wall thickness and clear passages improve performance.

Brackets: Bolt holes can be cast so that machining can be eliminated completely in some cases.

Other alloys

Steel: The process is not suitable for most steel castings because of the danger of carbon pick-up from the foam pattern. The carbon pick-up is not uniform, but carbon-rich 'inclusions' are found in carbon steel castings made using the process. Residues from the pyrolysis of the EPS become trapped in the liquid steel giving rise to areas of high carbon, up to 0.7%C, often under the upper surfaces of the casting where the EPS residue has been unable to float out of the steel. Such defects are unacceptable in most steel castings. Even the use of foam patterns made from PMMA does not entirely eliminate the problem. Some high carbon steels, such as austenitic manganese steel which has around 1.25%C content, can be successfully cast using lost foam.

Wear resistant castings: Elimination of flash and dimensional precision give LF a major advantage over other casting methods. Grinding balls can be made in enormous clusters. Shot blast parts are made in large numbers. Slurry pump bodies can be cast with minimum machining saving many hours of machining time. Wherever the rather high tooling costs can be justified, LF has become the standard method of manufacture.

Duplex castings: The process lends itself to making castings containing inserts, such as tungsten carbide or ceramic. The inserts are fitted into the foam pattern before casting.

The future

LF casting has now passed through the critical development stage to become a mature foundry process. While many of the early technical problems have been overcome there are still some to be solved:

Cluster assembly, still requires too much manual work.
Methoding, still a matter of trial and error, though the current work on modelling is helping.
Distortion of foam patterns can be a problem, but more and more the coating is seen as a way of stiffening fragile patterns.
Glueing, expensive and not easy on non-flat joint lines.
Automation, better methods of mechanically handling foam patterns for cluster assembly and coating are needed.

Lost foam will not replace conventional casting processes. It is seen as a clean process, all the used sand can be easily reclaimed and chemical residues burnt off so that it is probably the most environmentally acceptable foundry process at present available. This will prove to be a growing asset in the future. But that is not the main reason. For the major automotive groups the reason is the design flexibility offered.

Chapter 16

Coatings for moulds and cores

The need for a coating

When liquid metal is cast into a sand mould or against a core, there may be a physical effect and a chemical reaction at the sand/metal interface. Either may result in surface defects on the finished casting.

Physical effects – metal penetration

Liquid metal penetrates the pores of the sand mould or core giving a rough surface to the casting. The degree of penetration is dependent on:

metallostatic pressure
    penetration is most severe in the lower parts of the casting and with high density metals such as iron and steel
dynamic pressure
    most severe where the metal stream impacts on the mould or core
pore size of mould or core surface
    open pores arise from the use of coarse sand or sand with poor grading, or from poorly compacted sand, due to imperfect core blowing or the use of sand having a high viscosity binder or one which has exceeded its correct work time
sand expansion
    stresses can form in the bonded sand due to differential thermal expansion, this can lead to the formation of mould or core surface cracks allowing ingress of molten metal (finning or veining).

Chemical effects
(1) Burn-on
A chemical reaction occurs between sand and metal. It occurs with

impure sand
    some impurities, particularly alkalis, reduce the refractoriness of the sand
some binders
    mostly those based on sodium silicate which have poor refractoriness so that liquid phases are produced at temperatures as low as 900°C

certain cast metals
   notably manganese steel which form oxides that react with silica sand
   to form low melting point liquids.
(2) Carbonaceous defects
   Organic mould and core binders degrade at high metal pouring
   temperatures, forming carbon bearing gases which can lead to carbon
   pick-up or surface pock marks due to lustrous carbon.

Metal penetration, sand burn-on and carbon defects occur most commonly
with ferrous metals, due to their high casting temperature, high density and
the reactivity of their oxides to silica sand. Conversely, moulds and cores for
aluminium alloy castings rarely display metal penetration and burn-on so
coatings are often un-necessary.

A refractory coating on the mould or core can reduce or eliminate metal
penetration and sand burn-on. For many years, foundries made their own
coatings, usually 'blackings', based on coke dust or graphite. The technology
of coatings has now become highly developed, and most foundries now
purchase coatings from specialist suppliers.

It is important to recognise the potential savings in shotblasting and
cleaning costs that can be made by using a good quality, well applied coating.

Coating selection

An effective coating should have the following characteristics:

   Sufficiently refractory properties to resist the metal being poured.
   Good adhesion to the substrate.
   No blistering, cracking or scaling tendency on drying.
   Good suspension and remixing properties.
   Good stability in storage.
   Good covering power.
   Good application properties.

Choice of coating and form of supply

When choosing a coating, consideration must be given to:

   Application method and equipment available
   Carrier – water or spirit-based
   Method of drying
   Performance required
   Mode of supply – tanker, tote tank, drums or powder.

Coatings are supplied in dry or liquid form.

Powder coatings

The attraction of a powder coating is that it costs less to transport than a liquid, it is easy to store and can reduce waste since precise quantities of the mixed coating can be made to order. The characteristics of a coating prepared from a powder pre-mix are highly dependent upon the mixing equipment used and the procedure adopted. If simple paddle mixers are used, long mixing times followed by ageing up to 24 hours may be required to develop the correct suspension and rheology. It is preferable to use special high shear mixers to develop the liquid coating and avoid the need for ageing. Suppliers of powder coatings will advise on the best mixer to use and on the correct testing procedure to ensure that the optimum coating properties are achieved.

Creams and ready-for-use coatings

A 'ready-for-use' paint is the easiest to use. So-called ready-for-use paints may also be supplied as light creams or slurries which require a small water addition to allow the user to control the final consistency of the coating to suit his particular application. During transport and storage of ready-for-use coatings, their is always some separation from suspension of the refractory filler, often a high density mineral, but such segregates are usually easily re-incorporated by simple stirring.

Components of a coating

Foundry coatings consist of:

  Refractory filler
  Liquid carrier
  Binder
  Rheology control system.

The refractory filler

This is the most important constituent since its properties decide the efficiency of the coating. The filler may be either a single material or a blend selected for specific applications. Fillers are chosen for their particle size and shape, sintering point, melting point, thermal conductivity, thermal expansion and reactivity towards the metal being cast and the moulding material on which it is being used. Zircon, being highly refractory, is used for the most demanding applications such as steel or iron castings where section thickness is great. Graphite and coke are commonly used for lighter iron castings, often in combination with zircon, silica or talc. Elimination or reduction of finning can often be achieved through the inclusion of lamellar platelet fillers such as talc and mica. A combination of silica and iron oxide is frequently used as an inexpensive refractory blend for cores for automotive castings. Graphite may be included to improve strip.

The liquid carrier

The liquid carrier is the vehicle for the total coating composition and serves to transport the filler onto the sand substrate. It must be removed before casting takes place. The most commonly used carriers are water and isopropanol, although other solvents such as methanol, ethanol, hydrocarbons and chlorinated hydrocarbons have also been used. Water is cheap and readily available but drying in an oven is usually necessary to remove it before casting. Foseco has introduced Rapid Drying Technology (RDT) to many of its water based coatings, shortening the drying time needed. Oven drying can be a problem on large moulds or cores. Isopropanol dries faster than water and can be removed quickly by burning so it is frequently used for large moulds and cores, but it is a more expensive carrier than water. Methanol and ethanol have higher vapour pressures at 20°C than isopropanol (Table 16.1) and dry faster still. Chlorinated hydrocarbons were at one time introduced for fast air drying, but they are now obsolete due to environmental concerns. The use of water based coatings has increased in recent years, also for environmental reasons.

Table 16.1 Alcohols used as carrier liquids

	Methanol	*Ethanol*	*Isopropanol*
Flash point (°C)	11	12	12
Boiling point (°C)	65.2	78.5	82.3
Vapour pressure at 20°C mm Hg	94	43	32
Explosion limit (Vol. %)	5.5–26.5	3.5–15	2.0–12
MAK value (ppm)	200	1000	400
Smell limit (ppm)	2000	350	100

The use of water based coatings on cores and moulds made with chemically bonded sand can effect the strength of the bond, and may give rise to casting defects due to surface friability of the core or mould. Silicate binders are particularly badly affected by water based coatings and spirit based coatings should be used instead. Cores made using phenolic-isocyanate resin processes can also be affected, and care must be taken to dry the cores quickly and immediately after coating.

The binder

The function of the binder is to bond together the filler particles and to provide adhesion to the mould or core. The binder often interacts with the whole of the coating composition and therefore cannot be considered in isolation.

The rheology control system

This provides the suspension system that prevents the filler from separating out during storage of the coating over extended periods. It ensures that the coating is homogeneous and ready for application with the minimum of agitation. It also controls the flow properties of the coating and is designed to suit the application method that is used. Coatings applied by dipping are different to those designed to be applied by brush, spray or overpouring and it is the rheology control system that affects whether the coating can be applied easily and evenly during use.

Application methods for coatings

The main methods of coating moulds and cores are:

   Dipping
   Brushing
   Swabbing
   Spraying
   Overpouring/flowcoating.

Each method needs different properties in the coating.

Dipping

This method involves submerging the core in the liquid coating, allowing it to pick up a suitable layer of the coating. It is the fastest method of application and is used by the majority of high production foundries. The core is dipped manually or mechanically and when it is removed from the coating bath, it must be allowed to drain to remove excess coating. The success of the operation depends on the consistency of the coating and its rheological properties. It is essential to ensure that a uniform thickness of coating is applied, that there is adequate wetting of the sand by the coating to give good adhesion and that excess coating drains freely from pockets in the core. Failure to do this will affect the dimensional accuracy of the resulting casting.

The type of liquid flow that is suitable for dipping is known as a pseudoplastic rheology. A pseudoplastic liquid experiences an immediate decrease in viscosity upon the application of shear. The original viscosity is regained as soon as the shear is removed. As the core is submerged in the coating, the shear involved reduces the viscosity to a low level, thereby ensuring effective coating of all areas of the core. Coating viscosity increases rapidly on removal of the core and drainage of excess coating ceases when the downward coating film weight is counteracted by the upward resistance of the gel. With a correctly designed coating, the necessary thickness is deposited in a very even manner without sagging, excessive dripping or tear formation.

Brushing

This is a simple application technique and one or more layers may be applied. The correct flow behaviour needed is known as thixotropic rheology (well known to users of household paints). The viscosity decreases slowly upon the action of shear and regains more quickly, but not immediately, on shear removal. The coating can be of almost jelly-like consistency in the can but will flow readily when brushed. It gives extremely good suspension of the filler but allows brush marks to flow out after application, leaving a smooth, even layer of coating on the core or mould.

Swabbing

This is similar to brushing, except that the implement used has long, soft bristles which tend to adhere to each other. It is easier to penetrate re-entrant areas of the mould with these long bristles and to apply a thick, smooth coating with the minimum of dripping. There is less tendency to mould damage or residual brush marks in the finished coating than with brush application. Both brushing and swabbing are rather slow and require a skilled operator to achieve the best results, but developments in coating rheology have significantly reduced the skill factor needed.

Spraying

This is a much faster technique than brushing but penetration and coverage of the sand is not as good as with brushing. Skill is required to ensure that deep pockets and re-entrant angles are fully coated. It is best used on moulds and cores having rather simple shapes.

Generally, sprayed coatings are of much lower viscosity than brushing or dipping varieties. They are applied under pressure from a spray gun delivering a spray of finely divided droplets on to the surface to be coated. A degree of thixotropy is desirable to prevent the coating from sagging and tearing. Correctly designed spray systems are available that give very good results, but older equipment may give problems as only very low viscosity coatings can be handled with these spray guns.

Overpouring/flow-coating

Overpouring, also known as flow-coating, is particularly suitable for larger cores and moulds, it consists of a pumping system and a specially designed nozzle. The coating is continuously pumped through the nozzle which directs it over the core, the excess runs back into a reservoir to be recirculated through the nozzle once more. Overpouring also requires special equipment for handling large moulds and cores. The rheology of the coating must be designed to allow an even layer build-up independent of the amount of overpoured coating used. The gel system should allow an even deposition of coating without the risk of residual runs and tear marks.

Mixing and testing suspensions

The suspension agents and rheology control agents used in coatings give the slurry some unusual properties which are difficult to measure in terms of simple viscosity. The viscosity of such slurries varies according to the shear forces applied. Users of household non-drip paints will be familiar with the way that the paint flows easily during brushing, then gels into a more viscous liquid as soon as the brush is removed. With coatings, a similar effect is seen. After shearing with a paddle mixer, for example, the slurry will take a little time (seconds or minutes) to recover its gel structure. This makes it difficult for the user, with only simple equipment, to check the consistency of the coating.

The usual method of checking coating consistency is by means of a Baumé reading, but the Baumé hydrometer must be carefully used to give consistent, meaningful readings.

The hydrometer must be clean.
Allow the mixed coating to settle for 5 minutes after mixing.
Gently immerse the hydrometer, supporting it manually until it finds the correct level (remembering that any shearing force on the slurry will alter its rheology).
Wash the hydrometer immediately after use, since dried-on coating is difficult to remove.
Always use the same type of hydrometer, coating suppliers will usually provide a suitable type.

Another method of testing coatings, particularly the less viscous slurries used for overpouring or spraying, is to measure the time taken for the slurry to pass through a 'flow cup'. There are a number of standard flow cups available, the 4 mm DIN cup is commonly used for coatings.

Neither of these test methods will measure the true coating properties of any slurry and, for dip-coats in particular, some foundries use a small test core which is weighed, dipped in the coating, dried, then weighed again to measure the 'dry weight pick-up'.

Equipment for mixing and applying coatings

Coatings applied by brush or swab are often used straight from the container in which they are supplied with perhaps a small adjustment in water or carrier liquid made and manually stirred in. For dip coating, spraying and overpouring it is usual to have a separate mixing tank in which the coating is adjusted to the required consistency and from which the dip tank, spray container or holding tank for flow coating can be topped-up when needed. Transfer between supplier's drums or containers to the mixing tank and from mixing tank to the working tank or vessel may be done by using a double diaphragm slurry pump.

Dip tanks should be fitted with a paddle stirrer to maintain the coating in suspension without excessive movement, which could affect the rheology of the coating. The temperature of the mixed coating should be maintained reasonably constant.

Equipment for the application of coatings by the overpour or flowcoat system consists of a method of pumping a continuous flow of coating over the core, allowing the surplus to run off the core to a catchment tray from which it is returned to the storage tank. Small portable systems may be used, in which a portable catchment tray is placed under the work and the coating pumped over the core through a coating gun by a diaphragm pump or by displacement with low pressure compressed air. Larger cores and moulds are usually coated at a fixed station having a coating storage tank of about 100 litres capacity allowing about 4 minutes uninterrupted coating time. Cranes or hoists and specially designed racks may be needed to support the work while it is being coated, with moulds being angled over the catchment tray in such a way that no pools of coating are left on the work.

Drying ovens

Cores coated with water based coatings must be thoroughly dried before casting. High production cores are usually passed through an oven on a conveyor. Warm air, infra-red and de-humidifying ovens may be used, in all cases good air circulation is needed to carry away the moisture laden air from the core surface. Air temperatures up to 100°C may be used. Large cores and moulds, which cannot be moved to an oven, may be dried using portable warm air blowers, or a gas torch.

Coatings for iron and steel foundries

Coatings can conveniently be divided into two types:

Coatings for high production cores
Coatings for jobbing cores and moulds.

Foseco supplies both types and following the merger of SMC with Foseco in 1998, the Foseco range of coatings has been supplemented by the SEMCO and TENO range of coatings.

Coatings for high production foundries

The RHEOTEC series of coatings

RHEOTEC coatings are water based coatings designed for dipping of cores which are then dried in a warm air oven. The coatings can be supplied as powders, creams or 'ready-for-use' slurries. They help to provide a clean 'as cast' surface finish on the internal passages of iron castings.

High production iron foundries use cores bonded with resin, usually phenolic-urethane or hot box. At the temperature of iron casting, the cores are prone to expansion defects which cause 'finning' or 'veining' defects in the castings. Veining defects arise from metal entering cracks in the core surface which form as a result of the thermal stresses generated by the expansion of silica sand during casting. In severe cases, the effect can lead to metal penetration rather than simply a vein defect.

Phenolic-urethane cold-box cores are more prone to veining defects than cores made by hot box, shell, ECOLOTEC, etc., due to the high packing density of the sand and the inherently low hot strength of the urethane binder. Castings suffering from veining require additional cleaning and grinding. If the veining or metal penetration occurs inside a water jacket or other internal cavity, it is impossible to remove and the performance of the casting, such as a cylinder head, block or manifold will suffer.

Iron castings made with cold box cores may also be affected by 'lustrous carbon' defects. The phenolic-urethane corebinder, at molten iron temperature, pyrolyses to form lustrous carbon on the molten metal front which affects its flow properties and may cause a 'wrinkled' surface on the casting which is unsightly and may be harmful. Core coatings can be designed to reduce veining and lustrous carbon as well as metal penetration and burn-on.

RHEOTEC XL anti-veining water based core coatings

The Foseco RHEOTEC XL coatings are particularly effective in reducing or even eliminating veining. They are used on cores for engine blocks, cylinder heads, ventilated disc brakes, manifolds, domestic boilers etc. where remedial cleaning is difficult or impossible. The coatings exhibit the following features:

High insulation (Fig. 16.1)
Excellent application characteristics
Controlled penetration of refractory fillers into the core substrate
High hot strength, giving a stable coating layer at casting temperature.

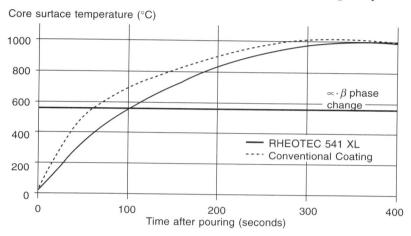

Figure 16.1   *Illustration of the superior insulating behaviour of RHEOTEC XL coatings.*

The reduced thermal shock experienced by the core when coated with RHEOTEC XL delays the α–β phase change of the silica sand, reducing the veining tendency. A single dip coat of RHEOTEC XL is sufficient. The coating is supplied either as a cream concentrate or in a ready-for-use consistency. The coating is applied at a suitable dilution to give a thickness of around 0.2–0.3 mm of dry coating layer. This is achieved at an application Baumé of 38–42.

RHEOTEC RDT fast-drying coatings

Water based coatings must be dried before the coated core or mould can be cast. Drying stoves use a combination of gas or electric heating (sometimes infra-red) with forced air movement sometimes with dehumidification to achieve the fastest drying at minimum fuel cost.

Refinement of the RHEOTEC system has led to the evolution of the RDT range of fast-drying RHEOTEC coatings. The technology behind these coatings is based on reducing the degree of 'wicking' of the coating into the core substrate. Drying times are significantly reduced compared with conventional coatings (Fig. 16.2).

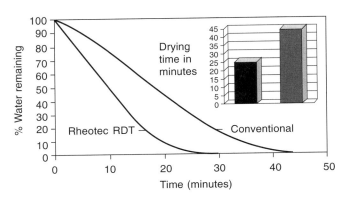

Figure 16.2 *Drying comparison, RHEOTEC RDT versus conventional coating.*

Typical water based dip coatings

RHEOTEC 102 — A general purpose, low cost coating for non-critical applications.

RHEOTEC 204 — Water based talc/graphite coating giving exceptional as cast surface finish, particularly good on cylinder heads and applications where veining is not a particular concern.

RHEOTEC 460 — Based on alumino silicate and iron oxide fillers, red in colour. Resistant to metal penetration and burn-on, good veining resistance. Suppresses lustrous carbon defects.

RHEOTEC 460 RDT — A faster drying version of 460. A special gel/surfactant system ensures optimum 'pick-up' with little penetration of water into the core surface, making it ideal for intricate cold box cores which suffer from strength problems when coated with water based products.

RHEOTEC 541XL — High performance water based dip coating. The special refractory technology provides superior anti-veining characteristics and extremely low levels of retained sand/coating residues after shake-out. Used in the production of complex iron castings such as cylinder heads and blocks, brake components and hydraulic castings where casting surface finish must be of the highest standard.

Solvent based dip coating

TENOSIL 301MPX — An alumino silicate solvent based coating used for thin section, complex urethane cold box cores where a water-based coating may cause problems due to strength reduction.

Coatings for jobbing moulds and cores

The HOLCOTE and SEMCO series of coatings

This range of coatings are ready-for-use water based coatings use a rheology control system designed for flow coating, spraying and brushing application for use on chemically bonded sand moulds and cores. Typical products in the range include:

SEMCO zir 35PX Zircon based coating designed particularly for flow coating.

HOLCOTE 110   Zircon based and suitable for steel castings and heavy section iron castings where organic sand binders such as furane or phenolic-urethane resins are employed. HOLCOTE is applied by brushing, swabbing, spraying and overpouring. Control of penetration into the sand can be obtained by adding up to 10% by weight of water. HOLCOTE coatings are thixotropic, the normally jelly-like material becomes mobile when shear forces are applied, as in stirring, brushing or spraying. After a short interval it will then recover its original gel properties. This means that HOLCOTE 110 will cling without dripping to a brush or swab, will liquefy, spread and penetrate when worked onto the mould or core surface and then level out, flattening out brush marks. HOLCOTE 110 is not suitable for dip applications and like other water based coatings is not recommended for silicate bonded moulds or cores. The applied coating can be cured by air drying, oven or infra-red drying. It can also be rapidly dried using a gas torch without any fear of blistering, lifting or cracking.

HOLCOTE 300 RDT   High performance zircon based coatings for steel and heavy iron castings. Can be applied by all methods, but especially dipping and overpouring. It has an enhanced drying capability.

Application guide: Dipping   75–85°   Baumé
                     Overpouring   73–84°   Baumé
                     Spraying   85–95°   Baumé
                     Brushing   85–100° Baumé

It will air dry in time but more positive methods are preferred. Even strong torching will not blister, crack or spall the coating.

Spirit based coatings

In cases where oven-drying of coated cores or moulds is not practicable, usually because the moulds or cores are too large to move, spirit based coatings are often used. After coating the mould or core, the coating is ignited and the carrier liquid burned off. Isopropanol is the most commonly used spirit carrier liquid, but ethanol and methanol may sometimes be added to improve the burning characteristics. The properties of carrier liquids are given in Table 16.1.

Isopropanol is chosen because of its low toxicity, low smell and low vapour pressure. Methanol is little used because of its potentially hazardous properties.

Spirit based coatings are usually supplied as ready-to-use coatings or heavy creams. Since spirit based coatings are used mostly on large moulds and cores for heavy castings, the fillers must be highly refractory and often contain a high proportion of zircon and other dense materials. This makes suspension difficult and sedimentation was always a problem, but developments in suspension and rheology control have improved greatly in recent years and sedimentation is no longer a serious problem.

Storage of spirit based coatings

Because they are required to burn out readily in use, spirit based pastes and slurries and the solvents used to dilute them are highly flammable and must be stored in flame-proof buildings according to local regulations such as the Petroleum Regulations in force in the UK.

Foseco spirit based coatings: ISOMOL and TENO product ranges

These coatings are used almost exclusively in jobbing foundries making iron and steel castings and can be used on any substrate i.e. CO_2-silicate, ester-silicate, furan etc. Application is by brush, spray or flow coating; curing is by air-drying or preferably by flaming-off.

TENO Coating ZN 75	a 'sulphur block' coating for ductile iron castings to prevent reversion of nodules to flake graphite at the surface of the casting. Designed for flow coating.
TENO Coating 70PX	based on alumino silicate for iron castings, designed for flow coating.
ISOMOL 210	based on zircon/graphite for heavy iron castings.
ISOMOL 300	based on zircon for steel castings, designed for brush application.
ISOMOL 380 FL ⎱ TENO ZBB P16 ⎰	based on zircon for steel and heavy section iron, specifically designed for flow coating.
ISOMOL 580 ⎱ TENOSIL MOPX4 ⎰	based on magnesite for manganese steel castings.

Evaporative polystyrene process coatings

In the evaporative casting, or lost foam casting process, a pattern of expanded polystyrene (EPS) is moulded to the required shape and attached to a running system also made of polystyrene. The assembly is coated with a refractory coating and after drying, is immersed in loose, unbonded sand which is vibrated to compact the sand around the pattern. Metal is poured into the pattern which is vaporised by the heat and replaced by the metal. The EPS shape is reproduced exactly by the metal, forming the casting (see Chapter 15).

The refractory coating plays an essential role in the process. If patterns are cast without a coating then sand burn-on occurs and severe erosion can take place near ingates and other areas where metal flow changes direction or speed. The coating must have the following properties:

It must provide an adequate refractory barrier to the molten metal.

It must have permeability, to allow liquids and vapours from the polystyrene to escape during mould filling.
It should have good strip and release from the finished casting.

In addition, the dried coating should have good strength, since it will then stiffen the pattern and reduce the tendency for it to distort during sand fill. Unlike other casting processes, the surface smoothness of the casting is not determined by the smoothness of the coating. It is the surface quality of the polystyrene pattern which dictates the casting surface finish, because the coating penetrates any crevices present in the pattern surface, causing the casting to replicate the pattern surface exactly, including any defects that may be present.

With iron castings, large volumes of gas are generated from the high temperature degradation of the pattern, so the permeability of the coating must be high, otherwise dangerous blow-back of metal up the sprue may occur. Care must be taken to ensure that only coatings designed specifically for the lost foam process are used. The STYROMOL and SEMCOPERM range of coatings has been developed specially for use with lost foam. All the coatings are water based and formulated for application by dipping. Grades are available for aluminium, grey and ductile irons and for special irons (p. 220). Patterns must be coated inside and out, since no cores are used in the process. Care must be taken to dry the patterns thoroughly after dipping, the drying temperature should not exceed 45°C to avoid thermal distortion of the foam pattern. Special de-humidifying ovens, with good air circulation are frequently used to ensure complete drying of the coating, both inside and out. Microwave heating has also been used to dry difficult inside surfaces, but it is necessary to remove most of the water by conventional warm air, then finish by microwave. Use of microwave heating alone will cause the water in the coating to boil before it dries, and this may cause bubbles or even cause part of the coating to flake away.

SEMCOPERM M66 has been developed for coating EPS patterns used for making iron castings by the lost foam process. It is particularly suitable for vacuum assisted casting. The permeability is such that metal pull-through is unlikely to take place. The coating is supplied as a heavy slurry at 95–100° Baumé. It must be diluted with water to 55–65° Baumé for application by dipping or overpour methods.

Coatings for the full-mould process

This process uses polystyrene patterns to make large one-off castings. The patterns are usually machined from large blocks of expanded polystyrene. After coating, the patterns are surrounded with a self-setting resin bonded sand to form a rigid mould. The process is widely used to make large, grey iron press-tool castings, sometimes many tonnes in weight, having metal sections as much as 100–150 mm thick. The coating in such cases must be capable of being applied in a thick layer, without cracking or peeling, to provide an adequate barrier to the metal. Foseco supplies STYROMOL 702FM coating, developed specially for this purpose. Spirit based coatings must not be used for this application, since burning off the coating would result in loss of the pattern before the sand mould was formed.

STYROMOL 702FM is a dense thixotropic water based slurry containing a blend of graphite and refractories, which can be applied by spray or brush. To facilitate application, a small amount of water can be added to give a Baumé of 90–100°. This will give a coating thickness of 1.5–3 mm. The most effective method of application is via a spray system. Once applied, the coating must be completely dried prior to ram-up and casting. The coating must be air dried using warm air circulation (but the temperature must not exceed 45°C or there is a risk that the polystyrene pattern may be damaged). It may take 2–3 days before all the moisture is removed.

The TRIBONOL process

Green sand moulding is the most widely used moulding process for the production of a wide variety of high production castings. Modern green sand moulding machines allow production rates of up to 300 moulds per hour, with ever increasing complexity of casting design. However, even with an optimised sand system and a high quality moulding machine, problems of poor surface finish due to metal penetration and burn-on persist in many cases. Casting surface finish problems from green sand moulds have traditionally been dealt with in a number of ways:

   Use of a high quality facing sand
   Use of a liquid based coating

Replacing problem areas of the mould face by a cored surface
Increased shotblast times.

Use of a liquid coating is often favoured, as it is easy to implement and involves low capital cost. However, liquid coatings are difficult to apply and slow down the production rate. Often they generate other problems such as reduced mould permeability and wet patches which can give rise to gas blow holes.

The TRIBONOL Process was developed to overcome the problems of wet coatings applied to greensand moulds. It has two components:

A dry, free flowing, highly refractory zircon-based powder coating called TRIBONOL
A special electrostatic delivery system for applying the TRIBONOL coating to a green sand mould.

Two types of delivery system are available, a manually operated 'hand gun' and a fully automated 'multi-gun' system for high productivity.

The TRIBONOL coating is given a frictional electric charge as it passes through the application gun. Since the charge is generated by friction, there is no electrical supply or high voltage involved. As the charged powder leaves the gun it is attracted to the green sand surface, which is at earth potential, and adheres to the surface. The spray pattern is not as directional as a liquid spray, because the electrostatic attraction effectively coats shadowed areas and deep pockets, depositing a very uniform layer of coating on the mould. Being completely solvent-free, there is no need to dry the coating, so that moulds can be immediately cored-up and rapid mould closure is possible with no loss of productivity.

The TRIBONOL Process is particularly suitable for repetition iron foundries, such as foundries producing engine blocks, cylinder heads, brake drums and other automotive components.

TRIBONOL ZF is an anti-pinholing coating designed to reduce nitrogen pinholing arising from the contamination of the moulding green sand with high nitrogen hot box and shell cores.

Miscellaneous coatings

HARDCOTE bond supplement

HARDCOTE 4 non-refractory, liquid dressing for hardening the face of greensand, silicate or resin bonded moulds and cores. Its function is to act as a supplementary bond holding mould faces, edges etc. in place in situations where friability might otherwise affect mould integrity. It is best applied by spraying but brush or swab may also be used.

HARDCOTE 4 should be left to dry in the air for around 10 minutes (for silicate or resin bonded moulds) and about 20 minutes for green sand. The moulds must not be closed before the coatings have dried.

Dressings to promote metallurgical changes

TELLURIT

These coatings contain metallic tellurium which acts as a chill-promoting medium for cast iron. They are used for producing a wear resistant surface layer on grey cast irons. The effect is localised to where the coating is applied, and a chilled layer about 3 mm deep is produced. TELLURIT can also be used on chills to enhance the chilling effect.

TELLURIT 2 is a paste for dilution with water to form a dressing of paint-like consistency. Coating is usually applied to moulds and cores which are still warm from drying or baking. The coating must be thoroughly dried before a second coat is applied.

TELLURIT 50 is diluted with isopropanol before use and either air dried or burned off before the mould is closed.

Tellurium vapour can be toxic and care must be taken to ensure that operators are not exposed to vapour either during drying the coating or during casting the moulds.

MOLDCOTE 50 is a flammable bismuth-containing, paste mould dressing for localised densening of cast iron. At molten iron temperatures bismuth dissolves in the iron and locally alters the solidification characteristics. Its effect on carbide stabilisation is less severe than tellurium resulting in a densening effect and avoiding chilling or retention of massive carbides. It can eliminate the use of metal chills and is useful for picking out isolated bosses, cylinder bores and other heat centres affected by open grain where conventional feed is difficult to apply. The paste is diluted with isopropyl alcohol and applied by brush to the area required. The dressing may be air dried or burned. Treated castings can be re-melted without fear of bismuth build-up since it is not retained during melting.

Other special dressings

CHILCOTE   A range of dressings for loose chills and cast-in metal inserts which form parts of castings.

CHILCOTE 4 is a refractory, self-drying dressing for external chills. It provides a permeable refractory layer on the chill face which permits lateral movement of any evolved gas, prevents welding and resists damping-back in closed green sand moulds awaiting casting. The chills are protected, preserved and more easily cleaned for reuse. It is used for coating external metal chills used in aluminium and zinc alloy castings and also on small chills for copper base alloys and cast iron.

CHILCOTE 10 is a fusion promoter. In the grey iron industry, it is sometimes desirable to include denseners and inserts which, after pouring, become part of the casting. Application of CHILCOTE 10 to the shot-blasted or pickled chill before inserting in the mould, ensures maximum fusion between metals.

SPUNCOTE 10 This is a specialist water based slurry coating, containing alumina as the refractory, formulated to provide a permeable coating with very low gas evolution for use in the centrifugal casting process for the manufacture of pipes and liners. The coating also assists traction on the coated face, allowing even flow of metal during the casting process and unrestricted extraction of the casting.

Coatings for foundry tools

Foundry tools used in the melting and handling of aluminium alloys (and, to a lesser extent, copper-base alloys) must be coated with refractory. The coating prevents the danger of iron contamination arising from the use of unprotected tools. HOLCOTE 110 is suitable. Plungers, skimmers, tongs etc. are cleaned and heated to 80–100°C and plunged into the HOLCOTE water based coating. The treatment may be repeated several times daily.

Iron and steel ladles and shanks are given three or four coatings once or twice daily before use. Several thin coatings give better protection than a single thick one.

The HOLCOTE coating must be thoroughly dried before being brought into contact with the liquid metal. This may be done by placing them near to the furnace for a period, then into the furnace flame immediately before being used.

FRACTON dressings are designed for the protection of troughs, launders, refractories etc., from attack by molten metal.

FRACTON 4 dressing provides a highly refractory top dressing to refractory work that is not wetted by molten metal or most slags, drosses and fluxes. The underlying material, brickwork, crucible or tool, is therefore protected and preserved. Potential build-up material does not stick but readily falls away. Skulls drop out cleanly and ladle and crucible cleaning is reduced to a minimum.

FRACTON 100A dressing is designed for the protection of metal launders, spouts, pig moulds etc. from attack by molten metal. The principal application is to cast iron launders used to convey molten metal for pipe-spinning processes. It may also be used to protect metal moulds etc. against attack by molten copper and copper alloys, nickel, aluminium etc. Its application to steel is limited by its carbonaceous nature.

Chapter 17

Filtration and the running and gating of iron castings

Introduction

The running and gating system carries out the following functions:

Controls the flow of metal into the mould cavity at the rate needed to avoid cold metal defects in the casting.
Avoids turbulence of metal entering the mould.
Prevents slag and dross present in the iron from entering the mould.
Avoids high velocity impingement of the metal stream onto cores or mould surfaces.
Encourages thermal gradients within the casting which help to produce sound castings.
Enables the casting to be separated from the running/gating system easily.

It is not possible to achieve all these requirements at the same time and some compromise is always necessary.

When considering running systems for iron castings, it is necessary to distinguish between grey iron and ductile iron. While some furnace slag may be present in liquid grey iron, it is not a dross-forming alloy so is not subject to inclusions due to oxidation of metal within the running system. Ductile iron, on the other hand, contains magnesium silicate and sulphide dross arising during treatment with magnesium. Moreover, residual magnesium in the treated liquid metal can oxidise when exposed to air to form more dross. Running system design must take this into account. The widespread use of ceramic foam filters in iron casting has enabled running system design to be simplified.

Conventional running systems without filters

The elements of a running/gating system for a horizontally parted mould are shown in Fig. 17.1.

Pouring bush The use of a properly designed pouring bush is recommended

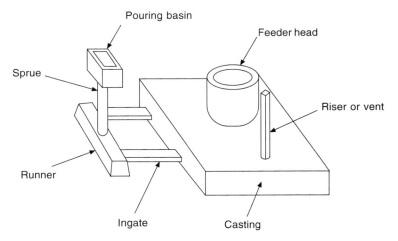

Figure 17.1   *The basic components of a running system.*

on all but the smallest of castings. The pouring bush should be designed in such a way so that the pourer can fill the sprue quickly and so maintain a near constant head of metal throughout the pour and retain most of the slag and dross within the bush. An off-set design incorporating a weir achieves this objective, Fig. 17.2. The pouring bush should be rectangular in shape so that the upward circulation during pouring will assist in dross removal. The exit from the pouring bush should be radiused and match up with the sprue entrance.

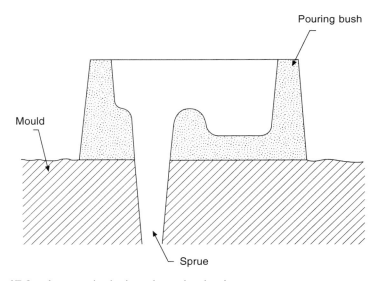

Figure 17.2   *A properly designed pouring bush.*

The practice of pouring directly down the sprue or the use of conical shaped bushes which direct flow straight down the sprue is discouraged as not only will air and dross be entrained and carried down into the system, but also the high velocity of the metal stream will result in excessive turbulence in the gating system.

Sprue The metal stream exiting the bush narrows in diameter as it falls and its velocity increases. To avoid air aspiration, the sprue should taper with the smaller area at the bottom, but in mechanised green sand moulding with horizontal parting, this is not possible since the sprue pattern must be tapered with the larger area at the bottom to allow the pattern to be drawn from the mould. Refractory 'strainer cores' may be sited at the base of the sprue to restrict the flow of metal, allowing the sprue to fill quickly and minimise air entrainment.

Sprue base Because stream velocity is at its maximum at the bottom of the sprue it is important that a sprue base be used to cushion the stream and allow the flow to change from vertical to horizontal with a minimum of turbulence. Recommended sizes of the sprue base are, a diameter two–three times the sprue exit diameter and depth equal to twice the depth of the runner bar.

Gating ratio This is the relationship between the cross-sectional area of the sprue, runner and gate. The system may be 'pressurised' or 'unpressurised'. A pressurised system is one in which the gates control the flow. A gating ratio of 1:1:0.7 is pressurised and is suitable for grey iron. An unpressurised system having gating ratio 1:2:3 is controlled by the sprue and is suitable for ductile iron since the reduced turbulence in the runners and gates limits the formation of dross.

Runners Runner cross-sections used in iron casting are usually rectangular (with some taper to allow for moulding), with width to depth ratio of 1:2, gates are taken from the bottom of the runner. It is presumed that the tall runner allows slag and dross to collect in the upper part of the runner. The distance between sprue and the first gate should be maximised for effective inclusion removal. The runner should extend beyond the last gate so that the first cold, slag-rich metal is trapped at the end of the runner.

Gates Ingates should ideally enter the mould cavity at the lowest possible level to avoid turbulence associated with the falling metal stream but practical moulding considerations often do not allow this. For grey iron castings, the ingates are usually thin and wide with a width to height ratio of about 1:4. The level of iron in the runner rises rapidly and is well above the top of the gate before iron flows through the gate so minimising entry of slag into the mould cavity. This shape of gate is easy to break so that grey iron castings are easily separated from the runners. The gate is usually notched close to the casting to break cleanly. One disadvantage of using gates to pressurise the system is that the velocity of the iron is high as it enters the mould and may cause erosion if the jet of metal impinges on core or mould wall.

Feeder head   This provides a reservoir of molten metal to compensate for any metal shrinkage occurring during solidification of the casting.

Riser   An opening leading from the mould cavity which relieves air pressure in the mould cavity as it fills with metal. It also acts as a flow-off, allowing cold or dirty metal to be removed from the mould cavity. If an open topped feeder head is used, a riser is not necessary.

Gating vertically parted moulds

Figure 17.3 shows examples of vertically parted running systems. Ideally the system shown in (a) may be considered best since each mould cavity is filled uniformly from below, whereas the top-gated system (b) allows metal to drop down the height of the mould with the possibility of turbulence and erosion.

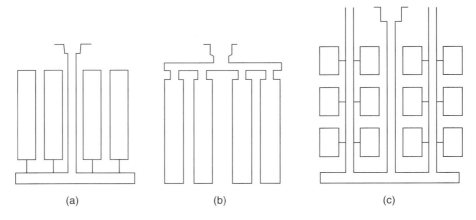

(a)   (b)   (c)

Figure 17.3   *Examples of vertically parted moulding systems: (a) A sprue/runner controlled system. (b) A runner/gate controlled system. (c) A multilevel system. (from Elliott, R.,* Cast Iron Technology, *1988, Butterworth-Heinemann, reproduced by permission of the publishers.)*

The effect of ingate size on filling time

In a pressurised running system where the gates control the metal flow, commonly used for grey iron castings, it is possible to calculate the total ingate area needed to fill a casting in a certain time. Grey iron at normal pouring temperatures is so fluid that small variations in temperature and carbon equivalent have little effect on the fluidity. The factors that control filling time are

the ingate area
the head of metal, for a bottom gated casting (Fig. 17.4a) this will be H at the start of pour and h at the finish;

the shape of the running system, in most cases the metal pours down the sprue, turns 90° into the runner bar and a further 90° into the ingates, slowing at each turn.

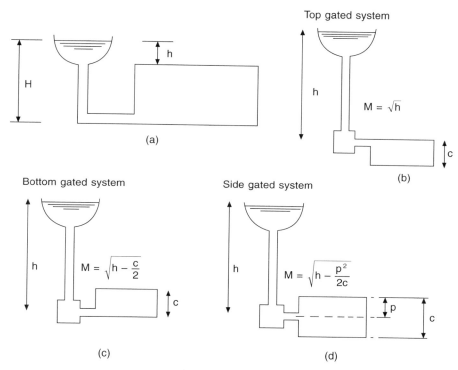

Figure 17.4 (a) The head of metal varies from H at the start of pour to h at the end. M is: (b) √h for a top gated system; (c) √(h – c/2) for a bottom gated system; (d) √(h – p²/2c) for a side gated system.

These factors are taken into account in the formula

$$A = \frac{8.12 \times W}{t \times M}$$

where A is total ingate area in cm^2
       t is pouring time in seconds, for casting and risers only
       W is the weight of the casting in kg
       M is related to the metal head (Fig. 17.4 b,c,d)

Example: A bottom gated casting of 25 kg is required to be poured in 15 seconds, h is 20 cm, c is 10 cm (Fig. 17.4)

$$M = \sqrt{(20 - 10/2)} = \sqrt{15} = 3.87$$

total gate area needed $= \dfrac{8.12 \times 25}{15 \times 3.87} = 3.5 \ cm^2$

This simple formula is remarkably accurate and useful as an initial guide.

Running and gating of iron castings is still the subject of controversy and although the above principles are widely accepted, rules are frequently broken and successful results still obtained. This is possible through improved melting and ladle practice which produces cleaner metal, improvements in the strength of green sand moulds, better cores and core coatings which resist erosion more than before. Above all, the widespread use of ceramic metal filters in running systems has not only eliminated most inclusions from the metal but has allowed more attention to be paid to increasing the utilisation of sand moulds and improving the yield of castings.

Filtration of iron castings

Filters were originally introduced to prevent non-metallic inclusions in the liquid metal from entering the casting. While this is still their main function, they are also used to simplify running systems allowing more castings to be made in a mould and improving the yield of castings.

Inclusions in iron castings

The occurrence of non-metallic inclusions in castings is one of the most widespread causes of casting defects encountered by the foundryman. The presence of these inclusions has a deleterious effect on cast surface finish, mechanical properties, machining characteristics and pressure tightness and can lead to the scrapping of castings.

There are two main categories of inclusions:
1. Inclusions which are generated outside the mould and carried in with the metal stream. The most common are:

   Melting furnace slag
   Ladle and launder refractories
   Ladle slag
   Flux residues
   Desulphurisation slag
   Alloying, nodularisation and inoculation reaction residues
   Oxidation products
   Contaminants and foreign objects.
2. Inclusions which are generated inside the mould. These include:

   Loose sand
   Mould and core erosion products
   Loose mould and core coating products
   Oxidation products generated in the running system
   Undissolved in-mould inoculants.

Traditionally, the incidence of inclusions has been controlled by:

The design of the ladle, e.g. tea-pot ladles
Design of the pouring basin
Gating system design, including slag traps spinners and whirl gates
Use of strainer cores (although these really only act as flow restricters to choke the base of the down-sprue to allow the sprue to fill quickly so that slag flotation occurs).

The use of filters is not a substitute for good melting and ladle practice but it can revolutionise running and gating practice.

Types of filter

For iron filtration there are three types of filter

Filter cloths
Ceramic foam filters
Cellular filters.

Filter cloths are made of woven refractory glass cloths. Although they reduce turbulence, they have a low filtering efficiency and a small open area (typically 25%) which acts as a severe choke in the running system.

Ceramic foam filters, Fig. 17.5 are the most efficient metal filters, they have an open pore, reticulated structure with a very high volume of porosity (over 90%) and a very high surface area to trap inclusions. The metal takes a tortuous path through the filter effecting the removal of very small inclusions by attraction and absorption to the internal ceramic pore surfaces.

Ceramic cellular filters have a 'honeycomb' structure with square section passages (Fig 17.6) and, since the ceramic walls are thin, can have up to 75% open area.

Ceramic metal filters work in several ways:

Coarse inclusions such as sand grains, large pieces of slag and dross films are trapped on the front face of the filter.
After some metal has passed through the filter, a 'cake' of material forms on its entry face which filters out finer particles. As the 'cake' builds up, it reduces the flow of metal through the filter so that there is a limit to the volume of metal that a particular size of filter can pass.
In addition to the physical filtration effect, there is a chemical attraction between the inclusions and the ceramic of the filter causing small inclusions to be trapped on the internal ceramic pore surfaces.
Finally, the smooth, non-turbulent metal flow through the filter reduces the exposure of fresh metal to air, limiting oxide film formation.

Ceramic filters were first developed for the aluminium casting industry, for use at temperatures up to about 900°C. Later, higher duty ceramics and

Figure 17.5   *Ceramic foam filters.*

Figure 17.6   *Cellular ceramic filters.*

improved manufacturing techniques allowed their use to be extended to a maximum temperature of about 1500°C, so that they could be used for copper based castings and in iron foundries. The development of ceramic materials suitable for the filtration of steel castings proved more difficult, mainly because the rather low superheat associated with carbon and low alloy steels caused freeze-off problems. This has now been overcome and ceramic filters are widely used in steel sand foundries and investment casting foundries for metal temperatures up to 1700°C. See Chapter 18.

Since the early 1980s, growth in the use of filters in the foundry industry has been spectacular. Metal filtration has been responsible for major improvements in the general quality of castings as well as significant increases in yield.

The benefits of filtration of iron castings

The obvious reason for filtering iron castings is to remove particulate inclusions. Other benefits can then be obtained such as increased yield, better machinability and better properties. The types of inclusions vary in different types of iron.

Grey iron

The commonly found inclusions in grey iron are:

(1) Sand inclusions from loose moulding sand or sand erosion. Loose sand lying in the running system will be trapped by an efficient filter but sand lying in the mould cavity will not be affected by a filter. A filter is not a substitute for proper sand practice.
(2) Slag arising from the cupola or furnace, ladle treatment slag, refractories or ladle surface dross from oxidation of the metal. Casting defects from slag material often are revealed only after machining, they are frequently accompanied by gas holes. The slag material usually consists of calcium silicates and oxides of iron, manganese, aluminium or other metals. Cupola slag is rich in silica and contains CaO from limestone.
(3) When coreless induction furnaces are used, oxides of silicon, manganese etc. form and may be stirred into the melt by the turbulence in the furnace, so that they can be carried over into ladles and castings.
(4) Inoculant particles are sometimes found, particularly when cast mould-inoculation tablets are used.
(5) Manganese sulphide inclusions form in relatively high sulphur irons as the metal cools and solidifies. They float to cause cope surface gas holes but these inclusions are not removed by filtration, since they form after the metal has passed the filter.

Those grey iron castings which are highly machined such as disc brake rotors and cylinder blocks, have the greatest risk of having inclusion defects exposed by machining and they benefit most from filtration.

Malleable iron

Two types of inclusion are often found in malleable iron castings:

(1) Sand inclusions which occur in the same way as in grey iron.

(2) Slag, usually lime-silica slag and oxides. Because of the relatively high furnace and pouring temperatures used, the slag particles remain liquid and are difficult to filter out. Use of strainer cores is widespread in malleable foundries to assist the flotation of slag in the sprue and pouring cup.

Ductile iron

In addition to sand and slag inclusions similar to those found in grey iron, ductile iron castings suffer from dross arising from the magnesium treatment process. Dross defects are a major problem in ductile iron casting production. The dross consists of magnesium silicate films and magnesium sulphide particles.

Magnesium silicate is formed during the magnesium treatment by reaction between MgO and SiO_2. Where high silicon nodularisers are used such as MgFeSi, more magnesium silicate is formed than when using pure magnesium nodularisers. The magnesium sulphide is usually present as clouds of fine individual particles. The silicate dross films and MgS clouds are usually found together.

Dross defects in castings are caused by carry-over from the ladle but also form during the casting process itself, since any exposure of the liquid metal surface to air will allow formation of dross films. Turbulence in the casting process increases dross formation.

The severity of dross defects is influenced by:

Magnesium content – as residual magnesium increases, the dross increases.
The initial sulphur content – high initial sulphur increases dross.
The cerium content of the iron – cerium oxidises preferentially, reducing magnesium silicate formation.
Mould conditions – high moisture content increases dross.
Carbon and aluminium contents – both promote dross if increased.

Scrap castings

In the production of automotive ductile iron castings, scrap from inclusions is typically around 2% or more. Much of the scrap is only found after machining the casting. The cost of machining is often much higher than the cost of the unmachined casting, for example a ductile iron crankshaft approximately trebles in value after full machining. The additional cost of filtration thus becomes easy to justify, so the majority of ductile iron castings are now filtered.

Yield

Inclusions are a cause of lower yield, particularly in ductile and malleable castings. This is because the running system must be designed to trap and control slag and sand. Ductile iron running systems are designed for quiet mould filling and are long and deep to allow slags to float out. Typically, 20% of the melt output of a repetition ductile iron foundry is used to form sprue, runners and slag traps. Use of an efficient filter in the running system results in a valuable yield increase.

Productivity

The smaller running system possible with filters uses less pattern space so that an extra casting may often be produced in the same pattern area.

Physical properties

Certain properties, particularly fatigue strength, are strongly influenced by non-metallic inclusions. The effects of large inclusions such as dross films are severe in both ferritic and pearlitic ductile iron, because notches are formed at which failure can begin. Dross defects at as-cast surfaces will reduce fatigue life substantially, reducing the fatigue limit by about 20–30% (Table 17.1). This is of great significance for castings such as crankshafts which fail by fatigue.

Table 17.1   The effect of surface imperfections on the fatigue properties of ductile iron

No.	Fatigue limit (N/mm^2)	Endurance ratio	Fatigue strength reduction factor	Reduction in fatigue limit (%)	Description of surface
1	270	0.35	–	–	Sound, fully nodular graphite
2	220	0.32	1.23	19	Dross stringer and isolated areas of flake graphite
3	220	0.28	1.23	19	Fully nodular graphite dross defects
4	182	0.24	1.48	33	Fine flake graphite various dross defects

Machinability

The elimination of inclusions which are hard and abrasive means longer tool life in machine shops, measurements have shown that tool wear when turning grey iron is improved by more than 20% by filtration of the castings. Even greater improvements have been found when turning ductile iron. The improved surface cleanliness of filtered castings also means that machining allowances can be reduced, since a smaller cut is needed to remove surface defects.

SEDEX ceramic foam filters

SEDEX foundry filters are ceramic foam filters, with an open-pore reticulated structure, a very high volume of porosity (over 90%) and a very high surface area to trap inclusions (Fig. 17.5). This structure provides a highly efficient inclusion trapping system and gives smooth, non-turbulent mould filling so that dross generation from re-oxidation is minimised. In order to use SEDEX correctly it is necessary to understand how the filter works. There are three mechanisms of inclusion capture and control:

1. Large inclusions and filmy dross inclusions are trapped on the front, or 'active' face of the filter. After some time a cake of inclusions forms which itself collects inclusions and augments the filtering process. Eventually this cake is so heavy that flow is stopped – the filter has become blocked. Dirty metal causes early filter blockage, and dross-forming alloys such as ductile irons are much more prone to this effect.
2. Small inclusions, point inclusions and micro-inclusions which escape the preliminary filter layer penetrate into the centre of the SEDEX where the extensive surface area and complex flow of metal ensure their separation and adhesion onto the ceramic surface.
3. Metal leaving the SEDEX filter does so smoothly, without excessive turbulence which might create inclusions on the clean side of the filter.

These three highly effective mechanisms mean that the iron is virtually inclusion-free. Measurements have shown that inclusions as small as 2 micrometres are found trapped in used SEDEX filters.

Filter blockage

Eventually the collected inclusions can block the filter and stop the flow of metal. To prevent this happening the filter must be chosen correctly, in particular, the correct size of the filter must be chosen. Flow through a SEDEX filter, leading to eventual blockage is illustrated schematically in Fig. 17.7. SEDEX filters are supplied in a range of sizes and porosity to suit the type and quantity of iron to be filtered (Fig. 17.8).

How to apply SEDEX filters

Figures 17.9 and 17.10 illustrate the use of SEDEX filters in horizontally parted moulds. The filter area and size is dependent upon the quantity of iron to be filtered. The choice of filter porosity is influenced by the type of iron to be filtered (Table 17.2).

The running system should be designed according to Figs 17.9 and 17.10.

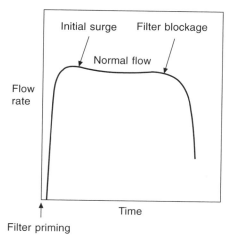

Figure 17.7 *Schematic pattern of flow through a SEDEX filter.*

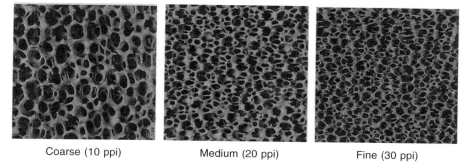

| Coarse (10 ppi) | Medium (20 ppi) | Fine (30 ppi) |

Figure 17.8 *SEDEX filters of differing porosity.*

Use the correct filter print to make the complete entry face of the filter available for filtration.

For grey iron the filter area should be at least two times and for ductile iron at least three times the downsprue area.

Position the filter as close as possible to the mould cavity.

The runner bar behind the filter should remain in the drag and be short, direct and turbulence free.

The use of Foseco filter prints is strongly recommended to ensure correct filter location (Figs 17.11, 17.12, 17.13 and 17.14).

For gating system calculation purposes the friction factor ξ is usually in the range 0.2 to 0.6 according to gating system and mould geometry.

The effective pouring height is determined by the relationship between cope height and ingate level.

SEDEX filter sizes and prints available are shown in Figs 17.11, 17.12, 17.13 and 17.14.

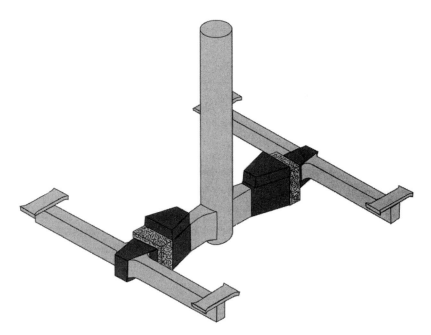

Figure 17.9 *The use of SEDEX filters in horizontally parted moulds.*

• Recommended gating system area ratios

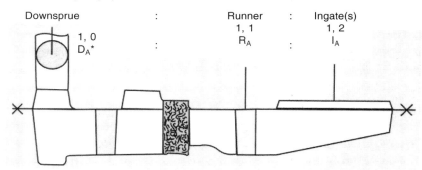

$$*D_A = \frac{22,6 \times W}{\xi \times \gamma \times t\sqrt{H}}$$

D_A : Downsprue area [cm²]
22,6 : Constant
W : Weight to be poured, including feeders (kg)
ξ : Friction factor (0.2 to 0.6 depending on gating system and mould geometry)
γ : Iron density [g/cm²]
t : Pouring time [s]
H : Effective pouring height [cm]

Figure 17.10 *Recommended gating system area ratios.*

Table 17.2   Choice of SEDEX filter porosity

Alloy or process	Porosity	Filtration capacity (kg/cm²)
Ductile iron	coarse (10 ppi)	1.4 to 2.0 max
Flake graphite iron	medium (20 ppi)	4.0
Malleable iron	fine (30 ppi)	4.0
Ductile Ni-resist	coarse (10 ppi)	0.8 to 1.0 max
Simo	coarse (10 ppi)	0.8 to 1.0 max
(Si-Mo ductile)		
Inmold process	coarse (10 ppi)	0.8 to 1.0 max

Note: The capacity of SEDEX ceramic filters is influenced by a variety of process factors so the values given above are for guidance only.

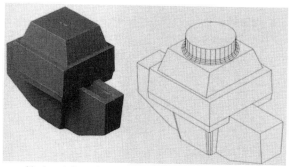

Horizontal position with 2 runner bars

Type and dimensions	Filter area (cm²)
2 - 50 × 50 × 22	25,00
2 - 50 × 75 × 22	37,50

Figure 17.11   *Filter print for horizontal position with two runner bars.*

For the application of SEDEX filters on DISAMATIC moulding lines, contact Foseco.

Cellular ceramic filters

A cellular ceramic is a refractory body which has been extruded into finely divided, individual channels called cells. The production method allows the external geometry of the filter and its internal geometry and density of the cells to be varied (Fig. 17.6). Foseco supplies CELTEX cellular filters in certain countries.

Cellular filters divide the metal stream into very small channels. Dross, slag and sand inclusions are collected on the filter face and along the internal filter cell walls. Analysis of the inclusions shows high concentrations of Mg, Al, Si and S in ductile iron applications, and Al and Si in grey iron applications. These examinations show that the impurities have a high affinity for the

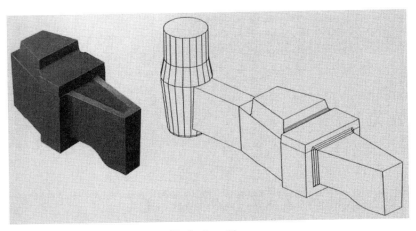

Vertical position

Type and dimensions	Filter area (cm²)	Type and dimensions	Filter Area (cm²)
4 – 40 × 40 × 15	16,00	4 – 35 × 50 × 22	17,50
4 – 50 × 50 × 15	25,00	4 – 50 × 50 × 22	25,00
4 – 60 × 60 × 15	36,00	4 – 60 × 60 × 22	36,00
		4 – 50 × 75 × 22	37,50
		4 – 50 × 100 × 22	50,00
		4 – 75 × 75 × 22	56,25
		4 – 75 × 100 × 22	75,00
		4 – 100 × 100 × 22	100,00
		4 – 150 × 100 × 22	150,00

Figure 17.12 *Filter print for vertical position.*

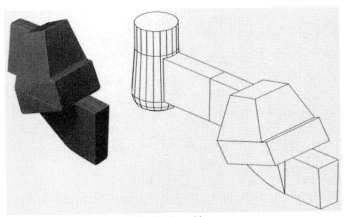

Inclined position

Type and dimensions	Filter area (cm²)	Type and dimensions	Filter area (cm²)
5 – 50 × 50 × 22	25,00	5 – 100 × 50 × 22	50,00
5 – 50 × 75 × 22	37,50	5 – 75 × 75 × 22	56,25
5 – 75 × 50 × 22	37,50	5 – 100 × 100 × 22	100,00
5 – 50 × 100 × 22	50,00	5 – 150 × 100 × 22	150,00

Figure 17.13 *Filter print for inclined position.*

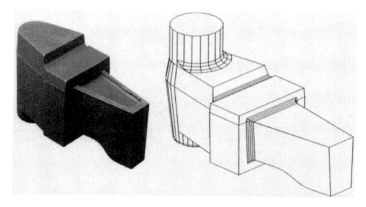

Vertical position combined with
downsprue base

Type and dimensions	Filter area (cm²)
6 – 35 × 50 × 22	17,50
6 – 50 × 50 × 22	25,00
6 – 50 × 75 × 22	37,50
6 – 75 × 50 × 22	37,50
6 – 75 × 75 × 22	56,25

Figure 17.14 *Filter print for vertical position with downsprue base.*

ceramic material, allowing particles much smaller than the cell openings to be trapped. Filtration occurs by two mechanisms, physically removing particles larger than the cell openings, and chemically attracting particles smaller than the cell opening.

The cellular filter is less effective than the equivalent SEDEX filter.

The filters are positioned in the gating system so that the metal must flow through the filter. A filter print is incorporated into the gating system pattern to form a cavity for the filter. Cellular filters have tightly controlled external dimensions, so that the filters accurately fit the print.

Choice of filter location follows the same principles as for ceramic foam filters. Figure 17.15 shows possible filter positions. To function effectively, the filter should not restrict the metal flow, to achieve this:

Filter frontal area = 4 to 6 times total choke area

This will achieve about the same pouring time as an unfiltered mould.

Flow rates

Metal composition, pouring temperature, metal head and filter position all have an effect on flow rate. The following data is representative of flow rates observed in practice.

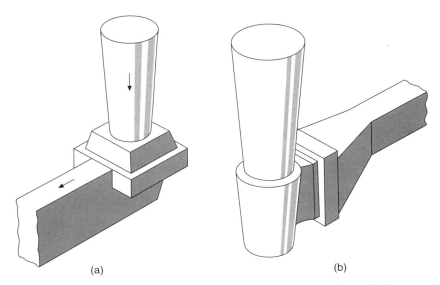

(a)   (b)

Figure 17.15   *Filter location for cellular ceramic filters. (a) Horizontal placement at sprue base. (b) Vertical placement at sprue base.*

Flow rate of iron through cellular filters

Filter description	Average flow rates			
	Grey iron		*Ductile iron*	
	(kg/s)	(lb/s)	(kg/s)	(lb/s)
2.0 × 2.0/100	5–6	10–12	3–4	6–8
3.0 × 3.0/100	10–14	24–30	7–9	15–20

Total poured weight

Non-metallic inclusions, collecting on the front face of a filter, can reduce the open frontal area of the filter through which metal will pass. In ductile iron applications the accumulation of dross and slag will eventually block the filter. The amount of ductile iron that a filter can pass before blocking depends on the foundry's practice, the location of the filter and the metal chemistry. The following guidelines have been developed from field trials.

Capacity of cellular filters

Filter description	Total flow of ductile iron before blockage	
	(kg)	(lb)
2.0 × 2.0/100	35–90	75–200
3.0 × 3.0/100	70–270	150–600

Filter blockage has not been observed in grey iron applications and the flow rate requirements determine the size and number of filters required.

Using filters to increase yield and number of castings per flask

Example 1: Grey iron cylinder block casting

The foundry wanted to change from 2 to 4 castings per mould (Fig. 17.16). Without a filter only two castings were possible because of the need for long runners to trap inclusions. With two SEDEX filters, a more compact running system was possible allowing four castings per mould with a 40% productivity improvement.

BEFORE USING SEDEX FILTERS

SEDEX: NONE
Pouring time: 14-16 sec.
Total Poured Weight:
125-130 kg

TWO UNITS DOCKING SYSTEM or SEPARATE SYSTEM

SEDEX (20 ppi): 2 pcs
Pouring time: 17-18 sec.
Total Pouring Weight:
Approx. 300 kg

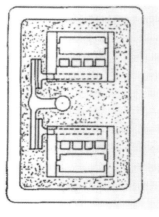 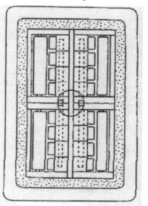

Result

Difference of defect ratio due to inclusion (dross/slag) before the use of SEDEX Filter and after the use of SEDEX filter.

- Repair ratio by welding:
    Decreased to 1/5
- Defect ratio after machining:
    Decreased to 1/10

By changing 2 Pcs/Flask to 4 Pcs/Flask, even though pouring yield is slightly reduced due to a larger pouring basin, the productivity is raised by about 40% – a significant improvement.

Figure 17.16 *Use of a SEDEX filter increases the number of cylinder block castings/ flask from 2 to 11.*

Example 2: Grey iron boiler casting

Use of SEDEX filters with a compact running system enabled two castings per mould to be made (Figs 17.17a,b).

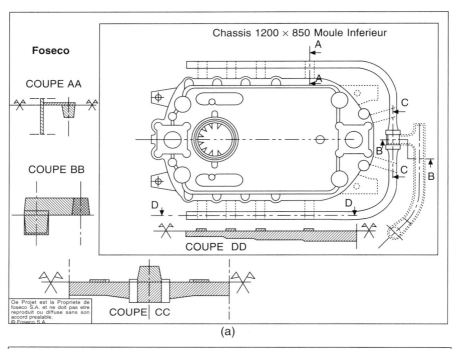

(a)

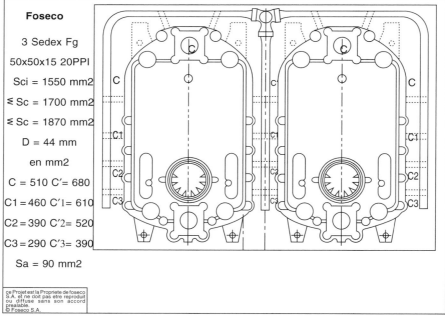

(b)

Figure 17.17   *Use of a SEDEX filter with compact running system to cast 2 boiler castings/flask: (a) one casting/flask; (b) two castings/flask.*

Using filters in vertically parted moulds

Example 1: Grey iron brake disc casting
Brake disc casting are frequently made on DISAMATIC moulding machines with vertically parted moulds. The disc castings are subsequently machined over a large part of their surface, any inclusions revealed by machining lead to both high tool wear and scrap castings. Filtration is clearly desirable but there are a number of problems of application:

Mould production rates are up to 350/hour.
Access to the open mould is not easy, filters must be placed by the coresetter machine to avoid slowing the mould rate.
Space to locate filter prints is limited.

Two systems of placing filters are possible (Fig. 17.18a,b,c).

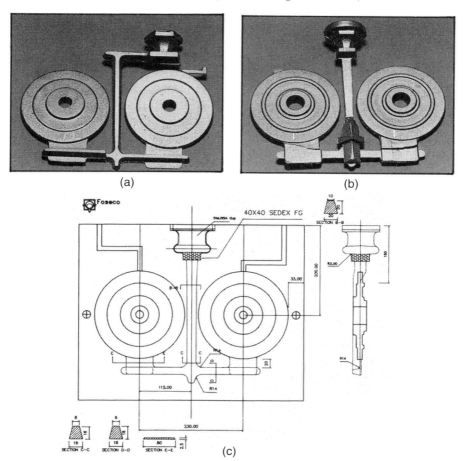

(a)   (b)

(c)

Figure 17.18 *Use of a SEDEX filter in vertically parted moulds. (a) unfiltered; (b) filter placed at base of sprue; (c) filter placed in pouring bush.*

Figure 17.18a shows the unfiltered running system, as is usual with DISAMATIC systems, the system is pressurised, being choked at the ingates. Figure 17.18b shows the SEDEX filter placed in a 'crush print' at the base of the sprue. In this position, the filter can be placed using the coresetter. The choke is prior to the filter. Figure 17.18c shows the filter placed in the pouring bush, here it is possible to place it by hand after the mould has been closed. The sprue base acts as choke.

The pour time is shorter with the SEDEX filter than without.

Running system	Pour time (s)
Unfiltered, choke at ingates	6.7
SEDEX filter at base of sprue	5.6
SEDEX filter in pouring bush	6.1

Both methods of using the filter are effective in removing inclusions.

Combined filter, feeder and pouring cup, the KALPUR direct pouring system

The concept of direct pouring into the top of a mould cavity has long been recognised as desirable, with the potential benefits of:

Improved yield
Simplified sprue, gating and feeding design
Reduced fettling costs.

Unfortunately, it was frequently found to introduce defects due to the turbulent flow of the metal in all but the simplest of castings. In addition, the impingement of high velocity metal streams caused erosion of moulds or cores. These objections can be overcome by pouring the metal through a ceramic foam filter situated at the base of an insulating pouring/feeding sleeve, the KALPUR unit. Clean metal, free from turbulence and oxide, fills the mould cavity and readily feeds the casting through the filter. Directional solidification and casting soundness is promoted and gates and sprues made unnecessary. The impingement problem is reduced because the metal velocity is reduced as it passes through the filter (Fig. 17.19).

Application to horizontally parted moulds

For manual moulding and simple moulding machines, the open pouring cup shape of Fig. 17.20 can be used; an example of its use is shown in Fig. 17.21. In horizontally parted automatic moulding lines, the KALPUR type shown in Fig. 17.22 can be used as in Fig. 17.23.

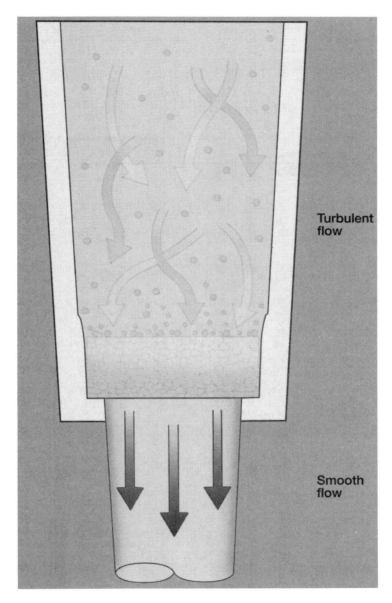

Figure 17.19 *A schematic view of the cleaning and flow-smoothing effect of pouring directly through a KALPUR unit.*

Figure 17.24 shows a ductile iron vice base, casting weight 26 kg made on a 20/24 Hunter moulding machine.

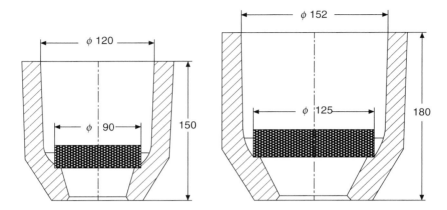

Figure 17.20   *Open KALPUR units for manual moulding. The units are supplied in a range of sizes.*

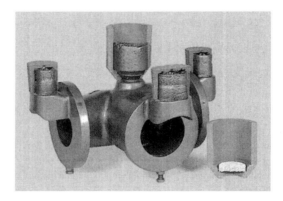

Figure 17.21   *Use of the open cup KALPUR unit.*

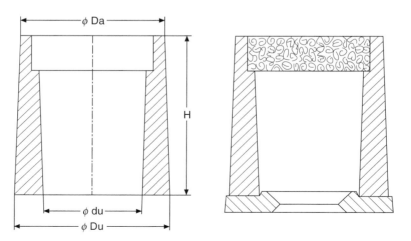

Figure 17.22   *KALPUR unit for horizontally parted automatic moulding lines.*

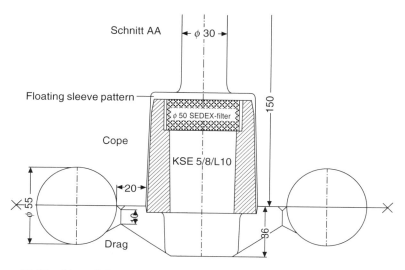

Figure 17.23 *Use of the KALPUR unit in horizontally parted automatic moulding lines.*

Figure 17.24 *Ductile iron vice base made on a Hunter machine using a KALPUR unit.*

	Before KALPUR	After KALPUR
Pour weight	34.5 kg	27.7 kg
Pouring time	14 s	8 s
Yield	69.3%	86.4%
Total scrap	20%	0.9%
Grinding time	2 min	1.5 min

Application of direct pouring to vertically parted moulds

Increasingly complicated grey and ductile iron automotive castings are being produced in vertical moulds for hydraulic, brake, suspension and transmission systems. Components such as conrods, hubs, brake drums, flywheels, brake discs, brake brackets and steering knuckles are now common production castings in vertical moulds. Since many of these are safety castings, they must usually be filtered. For ductile iron, feeders must also be incorporated so that sound castings can be obtained.

Figure 17.25a shows a conventional pattern layout for a ductile iron car hub casting made with three castings on the pattern plate. The running

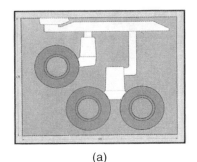

(a)

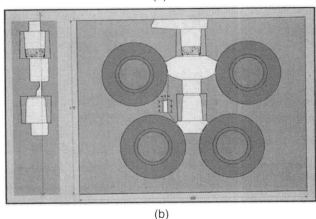

(b)

Figure 17.25 *Use of the KALPUR unit in vertically parted automatic moulding lines: (a) conventional layout; (b) using KALPUR.*

system allows mould erosion due to high metal velocity. Figure 17.25b shows a direct pouring system using a KALPUR direct pour unit at the top, and below it, a KALMIN S feeder sleeve ensuring correct feeding of the lower two castings. Casting yield increased from 61% to 78% and productivity by 33% while casting quality was improved.

Figure 17.26 shows a ductile iron differential housing, casting weight 30 kg made on a DISAMATIC 2070 moulding machine.

Figure 17.26 *Ductile iron differential housing made on a DISAMATIC 2070 moulding machine with a KALPUR unit.*

	Before KALPUR	After KALPUR
Pour weight	67 kg	44.5 kg
Moulding rate	215/h	230/h
Yield	44.1%	66.6%
Total scrap	8.5%	1.6%
Slag scrap	2.3%	0.02%
Grinding time	2.67 min	1.83 min

KALPUR direct pouring filter/feeder systems reduce mould and core erosion and ensure that there is hotter metal in the feeders than in the castings, giving ideal directional solidification from the casting to the feeder.

Chapter 18

Filtration and the running and gating of steel castings

Introduction

The running and gating system carries out the following functions:

Controls the flow of metal into the mould cavity at the rate needed to avoid cold metal defects in the casting

Avoids turbulence of metal entering the mould

Prevents slag and other inclusions from entering the mould

Avoids high velocity impingement of the metal stream onto cores or mould surfaces

Encourages thermal gradients within the casting which help to produce sound castings

Enables the casting to be separated from the running/gating system easily.

Controlling the flow of metal

Ideally the gating system should control the flow of metal into the mould. If lip pouring ladles are used, this can be readily achieved since the pourer is able to match the requirements of the gating system by altering the tilt of the ladle. It is more difficult with bottom pour ladles. The flow rate from a bottom pour nozzle/stopper rod system is determined by the nozzle size and the metal height in the ladle according to the formula:

$$t = \frac{4D(\sqrt{M_1} - \sqrt{(M_1 - \Delta M)})}{d^2 \sqrt{2g\rho\pi}}$$

where
t = pouring time (s)
D = mean ladle diameter (cm)
M_1 = initial weight of metal (kg)
ΔM = weight poured (kg)
d = nozzle diameter (cm)
g = acceleration due to gravity (981 cm/s^2)
ρ = density of liquid steel (7.7 g/ml)

(From C.S. Blackburn and B. Blair; Gating system design for bottom pour ladles, 35th SCRATA Conference, May 1992)

The flow rate varies considerably depending on the level of metal in the ladle. This is illustrated in Fig. 18.1 which shows the variation in pouring time of seven 750 kg castings from a ladle initially containing 5750 kg of steel. A gating system designed to accept the initial pouring rate of 750 kg in 18.23 seconds will not be suitable to accept the last 750 kg which is discharged from the ladle in 44.95 seconds, 2.5 times longer!

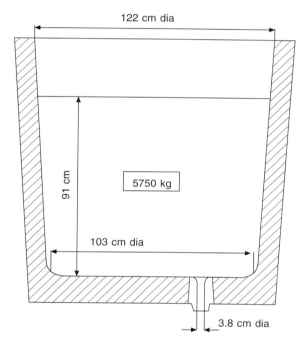

Weight in ladle (kg)	Pouring time (sec.)
5750	
5000	18.23
4250	19.68
3500	21.48
2750	23.96
2000	27.46
1250	33.37
500	44.95

Figure 18.1   *The variation in pouring time of seven 750 kg castings poured from a bottom-pour ladle initially containing 5750 kg of steel.*
(From Blackburn C.S. and Blair B. 35th SCRATA Conference, May 1992, courtesy CDC.)

The nozzle/stopper rod system of a bottom pour ladle is an excellent 'on/off' valve but ideally should not be used as a flow control valve. Attempts to use it to control flow result in breakup of the metal stream with consequent risk of reoxidation of the steel and possible erosion of the stopper rod itself. There is, however, no alternative if the gating system is to control the flow rate. A compromise must be reached for each cast.

Conventional running systems without filters

The elements of a running/gating system for a horizontally parted mould are shown in Fig. 17.1.

Pouring bush   The use of conical shaped bushes which direct flow straight down the sprue is discouraged as not only will air and dross be entrained and carried down into the system, but also the high velocity of the metal stream will result in excessive turbulence in the gating system. Incorrect alignment between nozzle and pouring cup will also cause metal splashing (Fig. 18.2).

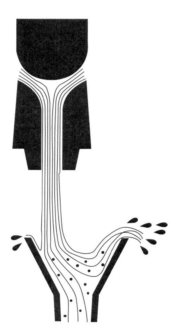

Figure 18.2   *Simulation of metal splashing due to incorrect alignment of nozzle and pouring cup.*
(From Blackburn C.S. and Blair B. 35th SCRATA Conference, May 1992, Courtesy CDC.)

A pouring bush designed as in Fig. 18.3 provides a larger target area, a shape which minimises splashing and sufficient volume to accommodate the maximum flow of metal obtained by opening the stopper rod fully thus reducing the need to throttle. The exit from the pouring bush should be radiused and match up with the sprue entrance.

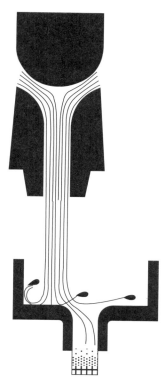

Figure 18.3 *Simulation showing the elimination of metal splashing due to optimisation of pouring cup design.*
(From Blackburn C.S. and Blair B. 35th SCRATA Conference, May 1992, Courtesy CDC.)

Sprue   The sprue should ideally be full of metal throughout the pour to avoid the possibility of slag being drawn into the mould cavity rather than being retained in the pouring cup. The metal stream exiting the bush narrows in diameter as it falls and its velocity increases. To avoid air aspiration, the sprue should taper with the smaller area at the bottom.

Sprue base   Because stream velocity is at its maximum at the bottom of the sprue it is important that a sprue base be used to cushion the stream and allow the flow to change from vertical to horizontal with a minimum of turbulence. Recommended sizes of the sprue base are, a diameter two–three times the sprue exit diameter and depth at least equal to the depth of the runner bar (Fig. 18.4).

Runner bar   Runner cross-sections used in steel casting may be square, rectangular (with some taper to allow for moulding), or circular (if refractory hollow-ware is used). It is presumed that a tall runner allows slag and dross

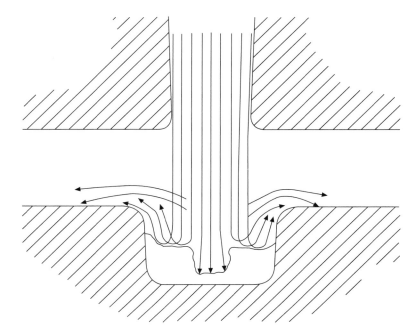

Figure 18.4   *Flat bottomed sump located at sprue base minimises turbulent metal flow.*
(From Blackburn C.S. and Blair B. 35th SCRATA Conference, May 1992, Courtesy CDC.)

to collect in the upper part of the runner. The distance between sprue and the first gate should be maximised for effective inclusion removal. The runner should extend beyond the last gate so that the first cold, slag-rich metal is trapped at the end of the runner.

Ingates   The ingate should introduce clean metal into the mould cavity with the minimum of turbulence. Low turbulence is best achieved by ensuring that the gating system is unpressurised by sizing the gates to have a total cross-section at least as large as that of the runner bar. The gating system must always be full which is achieved by taking ingates of the top of the runner bar and gating into the lowest part of the mould cavity. The ingate should also be sized so that its modulus is smaller than that of the section of the casting into which it enters. This prevents the formation of a small shrinkage cavity at the ingate/casting interface. If this cannot be achieved, the metal should be introduced into a feeder head.

Feeder head   This provides a reservoir of molten metal to compensate for any metal shrinkage occurring during solidification of the casting.

Riser   An opening leading from the mould cavity which relieves air pressure in the mould cavity as it fills with metal. It also acts as a flow-off, allowing cold or dirty metal to be removed from the mould cavity. If an open topped feeder head is used, a riser is not necessary.

Gating ratio   The gating system should be unpressurised with sprue : runner : ingate areas 1 : 2 : 2. If a bottom pour ladle is used, then the sprue size should be related to the nozzle size, with the nozzle bore controlling the flow although, as mentioned above, this is generally not possible because of the variable flow from a bottom pour ladle as the level in the ladle falls.

Gating of steel castings is not a precise science and each foundry develops its own preferred methods.

The use of ceramic foam filters

The primary function of the gating system is to introduce metal quickly into the mould without turbulence or non-metallic inclusions. Conventional gating design is always a compromise. When effective ceramic foam filters suitable for steel were developed, they led to immediate improvements in the quality of steel castings since, not only do filters allow inclusions in the steel to be eliminated more certainly than in the past, but they also reduce turbulence in the metal stream allowing the two main functions of a gating system to be readily achieved in a way that allows significant yield increases.

Inclusions in steel castings

The cleanliness of steel is, in the main, defined in terms of non-metallic inclusions. Improved properties of castings result from reducing the number of inclusions. Inclusions are introduced into steels during every stage of processing, from melting, trimming additions, deoxidation, tapping and teeming, as well as being generated within the mould itself. Inclusions are generally termed:

Exogenous – which arise from external sources and are typically particles of sand, refractory, moulding materials, melting and ladle slags and agglomerates of any of these.

Indigenous – which originate from chemical reaction between elements within the steel itself during melting and deoxidation, and reoxidation, which occurs during tapping and teeming. Such inclusions are essentially silicates, oxides, nitrides and sulphides, or more often complexes of these.

It has been estimated that 1 tonne of steel contains between 10^{12} and 10^{15} indigenous oxide inclusions. Each inclusion has associated with it a localised stress field, with inclusions of the order of 20 microns capable of nucleating fatigue cracks, and effecting reduction in fracture toughness and ductility.

A study of macro-inclusions in steel castings from 14 foundries in the USA found:

83% were reoxidation products which could be up to 5–10 mm in size
14% were from mould materials (sand and coating)
2% were slag
1% were refractory
1% were deoxidation products, always small in size, typically less than 15 μm

(Svoboda J.M. *et al. Trans. AFS* **95**. 187–202 (1987))

Reoxidation occurs during pouring from the ladle and within the running system as a result of turbulence, illustrating the importance of designing running and gating systems with care.

Effect of inclusions

Exogenous inclusions: Usually found in areas of rapid solidification, on upper cast surfaces or trapped under cores or adjacent to vertical walls. They can be surface or subsurface, revealed on machining and may also be associated with 'gassy' cavities. The size of these inclusions can vary from fine microparticles to gross macro inclusions up to about 75 mm diameter and 10 mm thick. Exogenous inclusions give:

Impaired surface finish.
Poor machinability.
Reduced mechanical properties.

Indigenous inclusions: are very varied and are predominantly of the following types:

Silicates – Usually glasslike inclusions which are relatively hard. Can have an adverse effect on fatigue and impact properties. Silicates have a deleterious effect on machinability by drastically reducing tool life.

Alumina – Primary source is from the deoxidation practice where aluminium is used as the principal deoxidiser. In general appearance the alumina crystals have a branched dendritic form. Alumina affects toughness and ductility, reduces the 'polishability' of surfaces and is abrasive to cutting tools. It also reduces fatigue strength. Aluminium is probably the most common deoxidant used in the steel industry, because of its strong deoxidising power and relatively low cost. It is normally added to the ladle on tapping.

Zirconia and calcium oxide – if Zr or Ca-containing deoxidant is used.

Sulphides – These are commonly manganese sulphide which can have different shapes, dependent on the conditions under which they form. Their morphology is dependent on the oxygen level associated with the steel and so is controlled by the amount of deoxidant, i.e. aluminium in the steel.

STELEX ZR ceramic foam filters

STELEX ZR filters (Fig. 18.5) are zirconia based foam filters designed to withstand the high temperatures encountered in steel casting. In sand moulds the filter is positioned in the running system using special filter prints designed by Foseco. STELEX ZR filters can also be used in precision casting processes such as investment casting where they are applied in specially designed pouring cups (Fig. 18.6). Ceramic foam filters are the most efficient metal filters, having an open pore, reticulated structure with a very high volume of porosity (over 90%) and a very high surface area to trap inclusions. The metal takes a tortuous path through the filter effecting the removal of very

Figure 18.5 *STELEX ZR ceramic foam filters.*

Figure 18.6   *STELEX ZR filter used in a ceramic pouring cup for investment casting.*

small inclusions by attraction and absorption to the internal ceramic pore spaces.

Ceramic metal filters work in several ways:

Coarse inclusions such as sand grains, large pieces of slag and dross films are trapped on the front face of the filter.

After some metal has passed through the filter, a 'cake' of material forms on its entry face which filters out finer particles. As the 'cake' builds up, it reduces the flow of metal through the filter so that there is a limit to the volume of metal that a particular size of filter can pass.

In addition to the physical filtration effect, there is a chemical attraction between the inclusions and the ceramic of the filter causing small inclusions to be trapped on the internal ceramic pore surfaces.

Finally, the smooth, non-turbulent metal flow through the filter reduces the exposure of fresh metal to air, limiting oxidation and reoxidation of the steel.

Ceramic filters were first developed for the aluminium casting industry, for use at temperatures up to about 900°C. Later, higher duty ceramics and improved manufacturing techniques allowed their use to be extended to a maximum temperature of about 1500°C, so that they could be used for copper based castings and in iron foundries. The development of ceramic materials suitable for the filtration of steel castings proved more difficult, mainly because the rather low superheat associated with carbon and low alloy steels caused freeze-off problems. This has been overcome with STELEX ZR filters which are widely used in steel sand foundries and investment casting foundries for metal temperatures up to 1700°C.

Since their development in the early 1980s, growth in the use of filters in the foundry industry has been spectacular. Metal filtration has been responsible for major improvements in the general quality of castings as well as significant increases in yield.

The effectiveness of STELEX ZR filters

STELEX ZR ceramic foam filters remove macro inclusions such as ladle slag, sand and furnace refractory eliminating the need for costly, time consuming defect removal and welding repair. Work-in-progress associated with castings awaiting rectification can be reduced significantly, resulting in on-time deliveries. Micro inclusions such as deoxidation and reoxidation products are removed by mechanical means at the surface of the STELEX ZR filter or by adherence to the internal filter structure. Mechanical properties and machinability of the castings are consequently enhanced and the need for excessive machining allowances can be eradicated. Figure 18.7 shows how filtration improves the cleanliness of an aluminium-killed steel.

Machinability tests

A programme of work has been undertaken at The Casting Development Centre to evaluate different methods for improving the machinability of casting. This work included a controlled experiment to measure the effect of ceramic foam filters. A standard test, ISO 3685 – tool life testing for single point turning tools, was used. Test bars of 100 mm diameter × 500 mm length were produced in carbon and low alloy steel using both filtered and unfiltered systems as shown in Fig. 18.8. The machinability of the test bars was compared by turning them using silicon carbide tipped tools at a cut depth of 2.5 mm until a flank wear of 381 μm was exhibited on the tool. The results are summarised in Table 18.1. This work demonstrates that the use of STELEX ZR filters can significantly increase both tool life and machine productivity. These results have been substantiated by production foundries, where improvements in the machinability of castings produced with STELEX ZR filters and also a reduction in tool wear and machine downtime have been achieved.

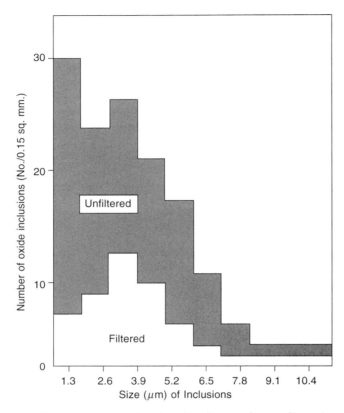

Figure 18.7 *Histogram comparing the cleanliness of an unfiltered and filtered Al-killed steel (sulphides not included in total number of inclusions).*

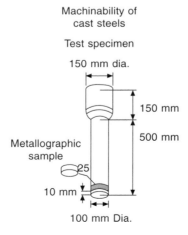

Figure 18.8 *Machinability test piece.*

Table 18.1   The effect of filtration of steel on machinability

Steel grade		Brinell hardness	Cutting speed (m/min.)	Cutting time (min.)	Metal removed (mm²)	Extension in tool life due to filtration
BS3100 A1 (carbon steel)	unfiltered	150	150	18.8	1 726 000	
			250	4.9	765 625	
	filtered	150	150	32.6	3 056 250	1.73 times
			250	13.0	2 031 250	2.67 times
BS3100 BT2 (low alloy steel)	unfiltered	280	150	13.2	1 331 250	
			250	2.0	312 510	
	filtered	280	150	17.3	1 621 875	1.22 times
			250	3.3	515 625	1.65 times

Flow control

A further function of STELEX ZR ceramic foam filters is to control metal flow and to facilitate non-turbulent filling of the mould. Many STELEX ZR filter users find this particular aspect of the filter's performance to be as beneficial as that of filtration; see Fig. 18.9. Non-turbulent flow reduces the risk of reoxidation of the metal as it enters the mould cavity, reduces lapping defects, promotes better surface finish, reduces air entrapment and offers the potential for faster pouring times.

The application of STELEX ZR filters to steel castings

For the successful application of STELEX ZR to steel castings, five factors need to be considered.

Support   The filter must be adequately supported in the running system.

Priming   The application technique and metal temperature at pouring must be suitable for good priming of the filter.

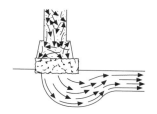

Figure 18.9   *STELEX ZR filters control metal flow, reducing turbulence.*

Filter capacity	The filter must be large enough to allow the casting cavity and feeders to fill before a filter blockage mechanism takes effect.
Metal flow rate	The filter must be large enough to allow the casting to fill in the optimum or required time.
Running system	The running system should be unpressurised

Support   The print the filter is positioned in must:

Adequately support the filter to ensure unnecessary stresses are not incurred that may damage the filter.
Prevent the possibility of metal bypassing the filter.
Be of a correct size to give a snug fit for the filter, while allowing for the minor dimensional tolerances relating to STELEX ZR.

A range of optimum resin filter prints and drawings are available for most sizes of square filters. The STLX FPI and FP7 prints should be used wherever possible and can be considered 'universal', a pictorial representation of these prints is shown in Fig. 18.10. Other filter prints should only be used where there are specific space restrictions and should be designed by Foseco.

Priming

It is important to ensure that the ladle temperature when the metal is poured onto the filter is at least 80°C above the liquidus temperature of the alloy. Typical minimum pouring temperatures are 1580°C for carbon steels and 1520°C for stainless steels. The filter should be positioned in a horizontal position at the base of the sprue, as characterised by the STLX FPI and STLX FP7 prints. With high alloy and stainless steel the level of superheat at pouring is usually high enough to ensure good priming. For carbon and low alloy steels extra temperature control and superheat may be required. It is preferable to use one large filter than multiple small filters, this is particularly important with carbon and low alloy steels.

Filter capacity

The amount of metal that can be passed through a STELEX filter depends on many factors, including pouring temperature, alloy composition, dirtiness of the metal and metallostatic pressure on the filter. An accurate prediction of filter capacity for a specific alloy in a specific foundry is difficult to make. If the metal is a high alloy or if a bottom pour ladle is used, high capacities will be achieved.

The deoxidation practice has a big impact on filter capacity. The use of calcium silicide and aluminium can be considered to be clean practice. Titanium and zirconium will dirty the metal and reduce the capacity of the filter. Tables 18.2 and 18.3 provide a guide to capacities that can be achieved.

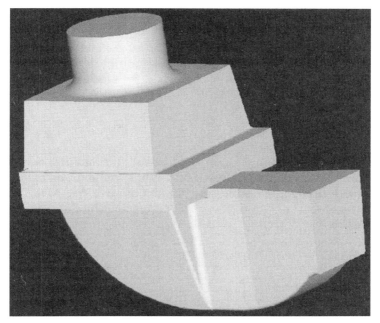

STLX FP1

STLX FP7

Figure 18.10 *Recommended STELEX filter prints.*

Metal flow rate

The same factors that affect filter capacity affect the metal flow rate through the filter. Tables 18.2 and 18.3 provide a guide to flow rates that can be achieved. Naturally, if the filter starts to become blocked, the flow rate will be reduced.

Running system design

The running system should be unpressurised after the filter and as simple as possible, long running systems and other methods of inclusion and flow control are not required and should be avoided. The filter area needs to be at least 4.5 times the sprue base area. The following ratios are recommended.

Sprue area	Filter area	Runner area	Ingate area
1	4.5 Minimum	1.1	1.2

Determination of the metal capacity and flow rates for STELEX filters

Selection of the correct STELEX filter for a given application is dependent on several factors:

Alloy   carbon/low alloy or high alloy/stainless ⎫
Ladle practice   lip pour or bottom pour   ⎬ factors which define fluidity/cleanliness
Deoxidation practice   Zr/Ti or CaSi/Al   ⎭

The amount of metal which must pass the filter
The mould fill time or metal flow rate required

Use Tables 18.2 and 18.3 as a guide to the selection of the correct filter. Note that if it is intended to apply more than one filter to a casting, special assistance from Foseco should be sought.

KALPUR ST direct pour unit

The KALPUR ST direct pour unit combines a filter and feeder sleeve into a single unit, Fig. 18.11. This unit replaces an existing feeder on a casting and then serves as a combined sprue, filter and feeder. Metal is introduced into

Table 18.2 Typical capacity of STELEX filters for carbon/low alloy steels

| Pouring method | Lip pour | | | | Bottom pour | | | |
| Deoxidation practice | Zr/Ti | | CaSi/Al | | Zr/Ti | | CaSi/Al | |
STELEX type	Capacity kg	Flow rate kg/sec	Capacity kg	Flow rate kg/sec	Capacity kg	Flow rate kg/sec	Capacity kg	Flow rate kg/sec
50 × 50 × 20	35	2.5	53	2.5	46	3.1	68	3.1
55 × 55 × 25	45	3.0	68	3.0	59	3.8	88	3.8
75 × 75 × 25	85	5.6	128	5.6	111	7.0	166	7.0
100 × 100 × 25	150	10.0	225	10.0	195	12.5	293	12.5
125 × 125 × 30	235	15.6	353	15.6	306	19.5	458	19.5
150 × 150 × 30	340	22.5	510	22.5	442	28.1	663	28.1
50 dia. × 25	30	2.0	45	2.0	39	2.5	59	2.5
60 dia. × 25	45	2.9	68	2.9	59	3.6	88	3.6
70 dia. × 25	60	3.8	90	3.8	78	4.8	117	4.8
90 dia. × 25	95	6.4	143	6.4	124	8.0	185	8.0
100 dia. × 25	120	7.8	180	7.8	156	9.8	234	9.8
125 dia. × 30	185	12.2	278	12.2	241	15.3	361	15.3
150 dia. × 30	265	17.7	398	17.7	345	22.1	517	22.1

Table 18.3 Typical capacity of STELEX filters for high alloy/stainless steels

Pouring method Deoxidation practice STELEX type	Lip pour				Bottom pour			
	Zr/Ti		CaSi/Al		Zr/Ti		CaSi/Al	
	Capacity kg	Flow rate kg/sec	Capacity kg	Flow rate kg/sec	Capacity kg	Flow rate kg/sec	Capacity kg	Flow rate kg/sec
50 × 50 × 20	46	3.5	69	3.5	61	4.4	102	4.4
55 × 55 × 25	59	4.2	88	4.2	79	5.3	132	5.3
75 × 75 × 25	111	7.8	166	7.8	149	9.8	249	9.8
100 × 100 × 25	195	14.0	292	14.0	263	17.5	439	17.5
125 × 125 × 30	306	21.8	459	21.8	412	27.3	687	27.3
150 × 150 × 30	442	31.5	663	31.5	597	39.4	995	39.4
50 dia. × 25	39	2.8	58	2.8	53	3.5	88	3.5
60 dia. × 25	59	4.1	88	4.1	79	5.1	132	5.1
70 dia. × 25	78	5.3	117	5.3	105	6.7	176	6.7
90 dia. × 25	124	9.0	186	9.0	167	11.2	278	11.2
100 dia. × 25	156	10.9	234	10.9	211	13.7	351	13.7
125 dia. × 30	241	17.1	361	17.1	325	21.4	541	21.4
150 dia. × 30	345	24.8	517	24.8	465	31.0	775	31.0

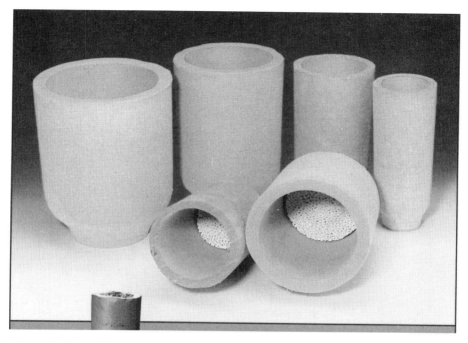

Figure 18.11 *KALPUR ST direct pour units.*

the mould cavity by pouring directly through the KALPUR ST unit, as illustrated in Fig. 18.12.

The sleeve performs initially as a pouring cup while the ceramic foam filter absorbs the impact energy of the metal stream, preventing sand erosion as the mould fills. As pouring progresses, the filter prevents non-metallic inclusions from entering the mould cavity and ensures non-turbulent flow. On completion of pouring, the KALPUR ST unit acts as an efficient feeder. The metal in the KALPUR ST unit is the last to be poured and, being the hottest metal in the casting system, provides ideal conditions for directional solidification. Benefits of using the KALPUR ST direct pour system include:

Elimination of the runner system, increasing yield, reducing cleaning and allowing more castings per flask

Efficient filtration

Low metal velocity in the mould, reducing erosion of sand and coating

Low turbulence of metal in mould reducing reoxidation

Efficient feeding through directional solidification

Lower casting temperature.

Computer solidification simulation

Extensive studies have been conducted to evaluate the thermal gradient

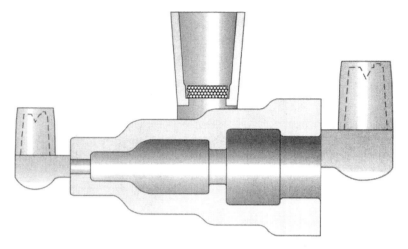

Figure 18.12   *The KALPUR ST unit is placed as close to the casting cavity as possible to preserve laminar flow from the filter.*

differences between conventional systems and direct pour practice. The examples shown in Fig. 18.13 illustrate the benefits obtained by the application of a KALPUR ST unit to a valve casting.

Productivity improvements

The KALPUR ST unit can be applied to a range of casting types. Impellers, for example, are usually cast with heavy running systems up into the centre of the casting body. A KALPUR ST unit can be applied as a central feeder, (see Fig. 18.14) and the metal poured directly into the casting. More castings can thereby be obtained from a ladle of metal. The promotion of excellent directional solidification has also resulted in the reduction in size or total removal of feeders on the perimeter of the casting (Fig. 18.15).

Selecting the proper size and positioning the KALPUR unit

There are two basic designs of unit, one for side feeder application and one for top feeder application, Fig. 18.16a, b. KALPUR ST units are supplied in a range of sizes (Fig. 18.11) having capacity from around 30 kg to over 250 kg of carbon or low alloy steel and more for high alloy steel (Table 18.4). Metal temperature, ladle practice (bottom versus lip pour), melting practice (cleanliness), metal deoxidation, and steel composition all determine the amount of metal that can be poured before filter blockage that would prevent proper pouring or adequate feeding after the mould is filled.

Factors resulting in low capacities and flow rates:

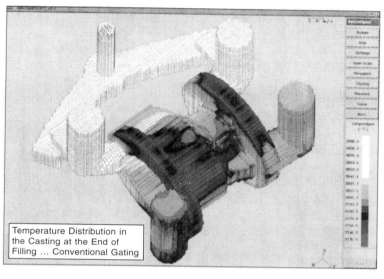

(a)

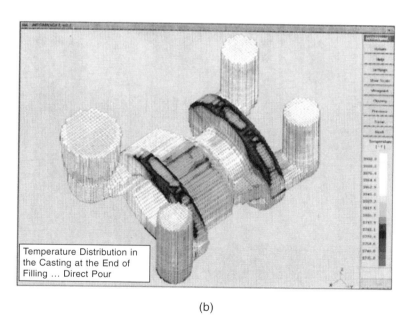

(b)

Figure 18.13 *(a) Temperature distribution within a valve casting immediately after filling through a conventional gating system. Large areas are much colder than the last metal poured. (b) Temperature distribution within a valve casting immediately after filling through a KALPUR unit shows higher overall temperature, illustrating reduced heat loss when direct pour units are used. (This figure is reproduced in colour plate section.)*

Figure 18.14 *KALPUR unit used as a central feeder.*

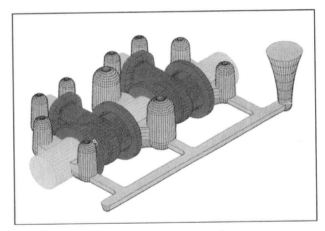

Conventional method

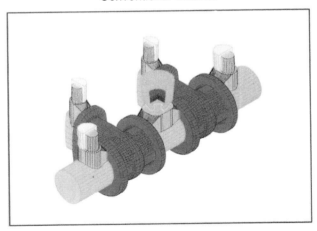

New KALPUR ST direct pour system

Figure 18.15 *The KALPUR ST direct pour system eliminates the conventional running system and reduces feeder requirements.*

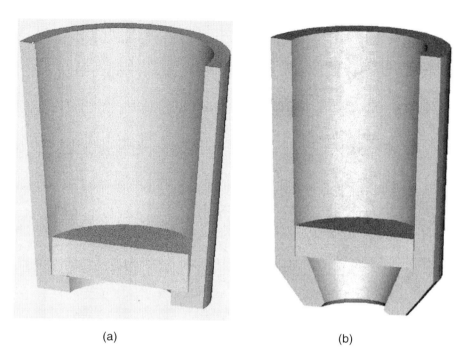

(a) (b)

Figure 18.16 *KALPUR ST unit: (a) for side feeder application; (b) for top feeder application.*

Table 18.4 Capacity and flow rates of KALPUR ST units

KALPUR size	Capacity range (kg)		Nominal flow rate (kg/s) (at 300 mm metallostatic pressure)	
	High level of deoxidation products	Low level of deoxidation products	Carbon steel	Stainless steel
KALPUR 50	30	90	2.0	3.0
KALPUR 75	65	200	4.4	6.6
KALPUR 100	120	355	7.8	11.7
KALPUR 125	185	550	12.2	18.4
KALPUR 150	265	795	17.7	26.5

Deoxidation with Zr, Ti
Molten metal containing large quantities of inclusion material
Low metal pouring temperatures
Low metallostatic pressure on filter.

A minimum pouring temperature of 1580°C for carbon steel and 1520°C for stainless steel is recommended to ensure priming.

The KALPUR ST system acts as an insulating feeder sleeve of similar size. The presence of a filter in a unit does not affect the feed efficiency. An efficient anti-piping compound should be applied to the KALPUR ST immediately after pouring is finished. Recommended materials are:

KALPUR ST 50 and 75　　　FERRUX 16
KALPUR ST 100, 125, 150　FERRUX 707F

The KALPUR ST system can incorporate a neck-down feature for reduced fettling and the possibility of knock-off.

To select the proper KALPUR ST unit for a given application, two criteria must be met: feeding capacity and filtration performance. First, one must determine which unit sleeve is large enough to satisfy the feeding requirement. Next, the filter in that unit must be evaluated for filtration capacity and flow rate. If the filtration capacity or flow rate is inadequate for the mould weight, cleanliness of the metal, or maximum pour time, a bigger KALPUR unit having a larger filter should be selected.

The KALPUR unit is rammed into position. It should be placed as close to the casting as possible and ideally should be located no more than 150 mm above the impact area (Fig. 18.12).

Cost savings through the use of STELEX and KALPUR

An analysis of steel casting cleaning costs has been undertaken by the Institut fur Giessereitechnik, Dusseldorf covering a number of steel foundries representing 10% of German steel casting production. The work showed that the cleaning costs associated with a casting can amount to 40% of its total manufacturing cost. Cleaning costs were further broken down as follows:

Reworking of casting defects (inclusions, penetration, shrinkage)	37%
Removal of feeding and gating systems	27%
Removal of rough surfaces (fins, flash and burn-on)	20%
Others	16%

Of these costs

52% were considered as avoidable
31% were considered as possibly avoidable
17% were defined as unavoidable

The use of STELEX ZR filters and KALPUR ST units enables major reductions to be made in the avoidable cleaning costs through simplification of gating and feeding systems and the removal of inclusions.

Chapter 19
Feeding of castings

Introduction

During the cooling and solidification of most metals and alloys, there is a reduction in the metal volume known as shrinkage. Unless measures are taken which recognise this phenomenon, the solidified casting will exhibit gross shrinkage porosity which can make it unsuitable for the purpose for which it was designed. To some extent, grey and ductile cast irons are exceptions, because the graphite formed on solidification expands and can compensate for the metal shrinkage. However, even with these alloys, measures may need to be taken to avoid shrinkage porosity.

To avoid shrinkage porosity, it is necessary to ensure that there is a sufficient supply of additional molten metal, available as the casting is solidifying, to fill the cavities that would otherwise form. This is known as 'feeding the casting' and the reservoir that supplies the feed metal is known as a 'feeder', 'feeder head' or a 'riser'. The feeder must be designed so that the feed metal is liquid at the time that it is needed, which means that the feeder must freeze later than the casting itself. The feeder must also contain sufficient volume of metal, liquid at the time it is required, to satisfy the shrinkage demands of the casting. Finally, since liquid metal from the feeder cannot reach for an indefinite distance into the casting, it follows that one feeder may only be capable of feeding part of the whole casting. The feeding distance must therefore be calculated to determine the number of feeders required to feed any given casting.

The application of the theory of heat transfer and solidification allows the calculation of minimum feeder dimensions for castings which ensures sound castings and maximum metal utilisation.

Natural feeders

Feeders moulded in the same material that forms the mould for the casting, usually sand, are known as natural feeders. As soon as the mould and feeder have been filled with molten metal, heat is lost through the feeder top and side surfaces and solidification of the feeder commences. A correctly dimensioned feeder in a sand mould has a characteristic solidification pattern: that for steel is shown in Fig. 19.1 the shrinkage cavity is in the form of a cone, the volume of which represents only about 14% of the original volume of the feeder, and some of this volume has been used to feed the feeder itself, so that in practice only about 10% of the original feeder volume is available to feed the casting. The remainder has to be removed from the casting as residual feeder metal and can only be used for re-melting.

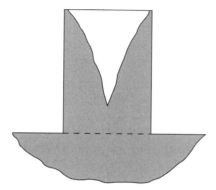

Figure 19.1 *Solidification pattern of a feeder for a steel casting (schematic).*

Aided feeders – feeding systems

If by the use of 'feeding systems' the rate of heat loss from the feeder can be slowed down relative to the casting, then the solidification of the feeder will be delayed and the volume of feed metal available will be increased. The time by which solidification is delayed is a measure of the efficiency of the feeding aid. The shape of the characteristic, conical, feeder shrinkage cavity will also change and in the ideal case, where all the heat from the feeder is lost only to the casting, a flat feeder solidification pattern will be obtained (Fig. 19.2). As much as 76% of an aided feeder is available for feeding the casting compared with only 10% for a natural sand feeder. This increased efficiency means greatly reduced feeder dimensions with the following advantages for the foundry:

A greater number of castings can be produced from the given weight of liquid metal
Smaller moulds can be used, saving on moulding sand costs
A reduction in the time needed to remove the feeder from the casting is possible
More castings can be mounted on the pattern plate and thus into the mould
Less metal is melted to produce a given volume of castings
Maximum casting weight potential is increased
Smaller feeders mean less fettling time and cost.

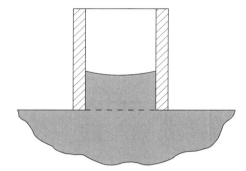

Figure 19.2 *Ideal feeder solidification pattern where all the heat from the feeder has been lost to the casting (schematic).*

Feeding systems

Side wall feeding aids are used to line the walls of the feeder cavity and so reduce the heat loss into the moulding material. For optimum feeding performance, it is also necessary to use top surface feeding aids. These are normally supplied in powder form and are referred to as anti-piping or hot-topping compounds. Figure 19.3 illustrates how the use of side wall and top surface feeding aids extends the solidification times. Now, however, suitable insulating discs (lids) are increasingly being used in place of hot-topping compounds for environmental reasons. For mass production castings 90% of feeders are closed or 'blind'.

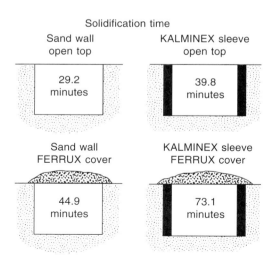

Figure 19.3 *Extension of solidification times with side wall and top surface feeding products for a steel cylinder 250 mm dia. and 200 mm high.*

Calculating the number of feeders – feeding distance

A compact casting can usually be fed by a single feeder. In many castings of complex design the shape is easily divided into obvious natural zones for feeding, each centred on a heavy casting section separated from the remainder of the casting by more restricted members. Each individual casting section can then be fed by a separately calculated feeder and the casting shape becomes the main factor which determines the number of feeders required.

In the case of many extended castings however, for example the rim of a gear wheel blank, the feeding range is the factor which limits the function of each feeder and the distance that a feeder can feed, the 'feeding distance', must be calculated.

The feeding distance from the outer edge of a feeder into a casting section consists of two components:

The end effect (E), produced by the rapid cooling caused by the presence of edges and corners.
An effect (A), produced because the proximity of the feeder retards freezing of the adjacent part of the casting (Fig. 19.4).

Where a casting requires more than one feeder the distance between feeders is measured from the edge of the feeder, not from its centre; and when the feeder is surrounded by a feeder sleeve the distance between feeders is measured from the outside diameter of the sleeve.

The effective distance between feeders can be increased by locating a chill against the casting mid-way between the two feeders (X1) and the natural end effect can be increased by locating a chill against the natural end (X). Chills should be of square or rectangular section with the thickness approximately half the thickness of the section being chilled.

There are therefore four possible situations:

Sections with natural end effect only (A+E)
Sections with natural end effect plus end chill (A+E+X)
Sections with no end effect (A)
Sections with no end effect plus chill ($A+X_1$)

Figure 19.4 shows the basis for calculating feeding distance in steel castings and all other ferrous alloys which freeze white, e.g. malleable and high alloy irons.

Ductile and grey iron castings

Alloy composition, casting section, mould materials and mould hardness all play a part in determining the actual feeding distance. The following tables are guidelines for green sand moulds having mould hardness 90° B scale, variations from these conditions will result in other feeding distances.

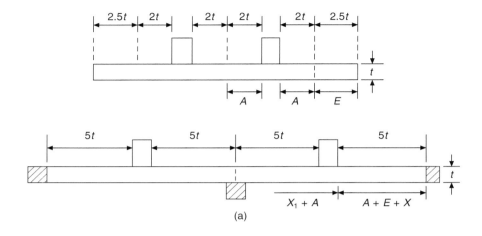

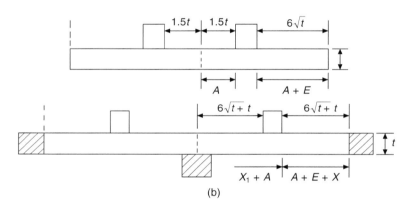

Figure 19.4　*Feeding distance in steel castings. (a) Plate (width: thickness >5:1. (b) Bar (width: thickness < 5:1).*

Table 19.1　Feeding distance factor (FD) for ductile iron castings

Carbon equivalent (%)	Feeding distance factor (FD)
4.1	6.0
4.2	6.5
4.4	7.0
4.6	9.0

Non-ferrous castings

Table 19.3 gives a feeding distance factor for some of the non-ferrous alloys; this factor is used in the calculation of an approximate feeding distance.

Table 19.2   Feeding distance factor (FD) for grey iron castings

Carbon equivalent (%)	Feeding distance factor (FD)
3.0	6.8
3.4	7.7
3.9	8.8
4.3	10.0

Table 19.3   Feeding distance factor for non-ferrous alloys

Casting alloy	Feeding distance factor (FD)
Al (99.99%)	2.50
Al 4.5% Cu	1.50
Al 7% Si	1.50
Al 12% Si	2.50
Cu pure	2.00
Cu 30% Ni	0.50
Brass	1.25
Al Bronze	1.25
Ni Al Bronze	0.50
Sn Bronze	0.75

The feeding distance in Fig. 19.4 is calculated using the feeding distance factor (FD) in the following manner:

Bars (width : thickness <5:1)	Plates (width : thickness >5 : 1)
$A = 1.5t \times FD$	$2t \times FD$
$E = (6\sqrt{t} - 1.5t) \times FD$	$2.5t \times FD$
$X = t \times FD$	$0.5t \times FD$
$X_1 = (6\sqrt{t} - 0.5t) \times FD$	$3t \times FD$

The calculation of feeder dimensions

The majority of foundrymen, even today, decide on feeder dimensions on the basis of experience. However, the application of calculation based on established theory and experimental data ensures the most efficient design of natural and aided feeders. In this section some guidance is given for calculating feeder dimensions from first principles whereas on page 334 there is a description of the various aids such as tables, nomograms, and computer programs developed by Foseco to make the determination of feeder dimensions much easier.

The modulus concept

Although this concept has some shortcomings it is, with the exception of computer programs, the most widely used acceptable and accurate method for calculating feeder dimensions.

The solidification time of a casting section is given by Chvorinov's rule:

$$t_c = kV_c^2/A_c^2 = kM_c^2 \qquad (1)$$

where

t_c is the solidification time of the casting section;
V_c is the volume of the casting section;
A_c is the surface area of the casting section actually in direct contact with the material of the mould;
k is a constant which is governed by the units of measurement being used, the thermal characteristics of the mould material and the properties of the alloy being cast.

M_c is the ratio of the volume of the casting section to its cooling surface area and is known as the casting's Geometric Modulus. It is expressed in units of length:

$$M_c = V_c/A_c \qquad (2)$$

The Modulus formulae for some common casting shapes are shown in Fig. 19.5.

Foundrymen, for the purpose of feeder size determination, are generally not directly interested in the exact solidification time but only that the feeder solidifies over a longer time than the casting. Having calculated the modulus of the casting section therefore, the modulus of the feeder is calculated as

$$M_F = 1.2 \times M_c \qquad (3)$$

where M_F is the modulus of the feeder required to feed a casting section having a modulus of M_c. This equation applies to natural feeders for most alloys. For grey and ductile iron castings because there is a graphite expansion phase during solidification the safety factor of 1.2 is considerably reduced.

The modulus extension factor (MEF)

The object of using feeding aids is to slow down the rate of heat loss from the surface of the feeder. It is possible to calculate how the improved thermal properties of the feeding aid compared with sand can reduce the feeder size.

From equation (1) the solidification time for a feeder is expressed as

$$t_F = kM_F^2 \qquad (4)$$

The constant k is composed of two parts:

the thermal characteristics of the mould material surrounding the whole of the feeder the properties of the metal within the feeder.

Shape	Dimensions	Modulus
(a) Cube	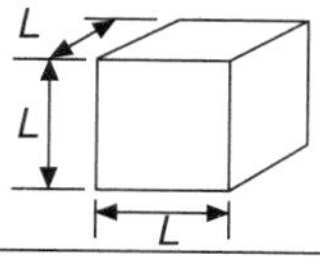 Side = L	$\dfrac{L}{6}$
(b) Cylinder	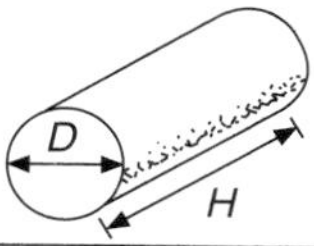 Diameter = D Height = H	$\dfrac{DH}{2D + 4H}$ *Note*: If $H = D$ the modulus is $\dfrac{D}{6}$
(c) Disc	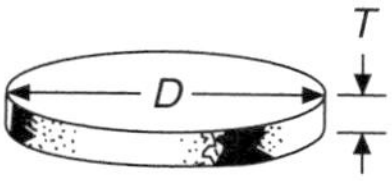 Diameter = D Thickness = T	$\dfrac{DT}{2D + 4T}$
(d) Bar or plate	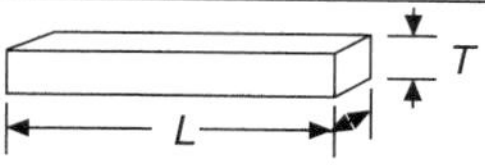 Length = L Width = W Thickness = T	$\dfrac{TWL}{2(TW + WL + LT)}$
(e) Endless cylinder (ends terminated by another part of casting)	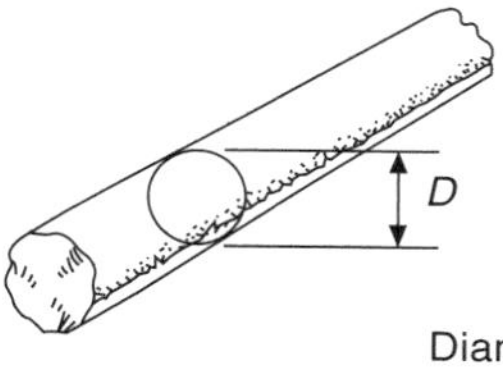 Diameter = D	$\dfrac{D}{4}$ *Note:* Because radial heat flow is faster than that from a flat surface, calculated moduli for endless cylinders may be reduced by multiplying by 0.85
(f) Endless plate (terminated on all sides by another part of casting)	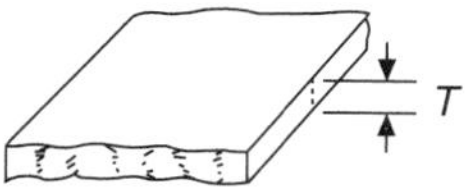 Thickness = T	$\dfrac{T}{2}$
(g) Endless bar (ends terminated by another part of casting)	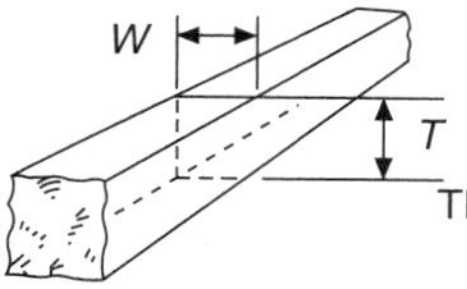 Thickness = T Width = W	$\dfrac{TW}{2(W + T)}$
(h) Endless hollow cylinder	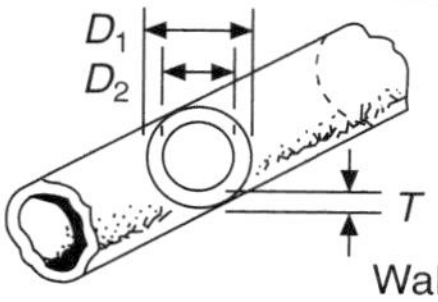 OD = D_1 Dia. core = D_2 Wall thickness = T	$\dfrac{D_1 - D_2}{4} = \dfrac{T}{2}$
(i) Annulus	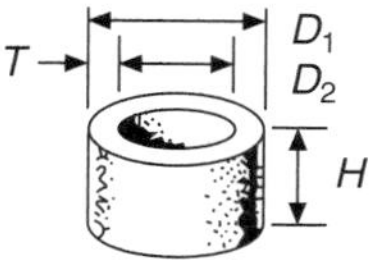 OD = D_1 Dia. core = D_2	$\dfrac{(D_1 - D_2)}{2(D_1 - D_2) + H}$ $= \dfrac{TH}{2(T + H)}$

Figure 19.5 *Modulus formulae for some common shapes.*

k can therefore be reduced to its two constituent components so that

$$k = cf^2$$

where

c is a constant depending on the properties of the metal being cast;
f^2 is a constant depending on the properties of the mould material.

There is no significance in the fact that this constant is expressed as a square other than mathematical convenience. Equation (4) can now be rewritten

$$t_F = C(fM_F)^2 \qquad (5)$$

The expression (fM_F) is known as the Apparent Modulus and f as the Modulus Extension Factor of the mould material surrounding all of the feeder's surfaces.

This approach provides a quantitative means of evaluating and comparing different feeding aids which should be designed to have the maximum possible Modulus Extension Factor compatible with the other properties necessary for a satisfactory product. In practice it is not customary to determine the absolute values of the constants relating solidification time to the modulus as these are seldom of interest.

Of greater concern is the improvement which can be expected from a variety of feeder lining materials when compared with the results obtained from the same size of feeder lined with the conventional moulding material – sand. For this purpose the Modulus Extension Factor (f) for sand is equated to unity and this serves as a basis for comparing other materials.

Example

Using this information it is possible to consider as an example a sound steel casting fed by one cylindrical feeder moulded in sand with a radius (r) of 16 cm and a height (h) of 32 cm. Because the feeder is attached to the steel casting, the bottom circular face of the feeder is a non-cooling face and the Geometric Modulus (M_F) of the feeder moulded in sand therefore is

$$M_F \text{ (sand)} = \frac{\pi r^2 h}{2\pi r h + \pi r^2} \qquad (6)$$

because in the example chosen $h = 32$ cm; $r = 16$ cm and $h = 2r$

$$M_F \text{ (sand)} = 2r/5 = 6.4 \text{ cm}$$

If in place of sand, a feeding aid system with a Modulus Extension Factor (f) of, for example 1.6 were to be used to line the feeder cavity and to cover the top surface of the molten steel in the feeder, then the aided feeder would remain liquid for the same time as the sand lined feeder if the Apparent Modulus of the aided feeder were to be equal to the Geometric Modulus of the sand lined feeder, i.e. for the equal solidification times:

$$M_F \text{ (sand)} = fM_F \text{(aided)} \qquad (7)$$

so that the Geometric Modulus of the aided feeder would be

$$M_F \text{ (aided)} = M_F \text{ (sand)}/f \tag{8}$$

or in the case chosen as the example

$$M_F \text{ (aided)} = 6.4/1.6 = 4.0 \text{ cm}$$

a cylinder where $h = 2r$ having a Geometric Modulus of 4.0 cm would be 20 cm diameter × 20 cm high.

The use of the feeding aid system quoted in this example therefore represents a reduction of approximately 75% in the feeder volume needed to achieve the same solidification time. The values for Modulus Extension Factors vary widely according to the type of feeding aid, the size and shape of the feeder sleeve and even the shape of the casting being fed. Foseco publishes tables showing the dimensions of sleeves available together with the Geometric and approximate Apparent Modulus of each sleeve.

Determination of feeding requirements

Steel, malleable iron, white irons, light alloys and copper based alloy castings

1. *Calculate casting and feeder modulus*
    (a) Divide the casting into sections and determine the important volume-to-surface ratios according to Fig. 19.5.

$$\text{Modulus} = \frac{\text{Volume}}{\text{Cooling surface}}$$

$$M_c = V_c/A_c \tag{9}$$

    (b) Determine the required feeder modulus (M_f) using the factor 1.2

$$M_F = M_c \times 1.2 \tag{10}$$

    (c) Provisional determination of the feeder from the feeder sleeve tables published by Foseco Companies.

2. *Calculate the feed volume requirement*
Feeders, whose dimensions obtained above satisfy modulus requirements do not necessarily always satisfy the total feed metal demand of the casting section. This must always be checked and if the feeder is found to contain insufficient available feed metal the feeder dimensions must be increased. Generally it is preferable to retain the feeder diameter dictated by modulus considerations and increase the feeder height. Sometimes it is more convenient to increase the feeder diameter in place of or as well as height. Never reduce the feeder diameter below that necessary to meet modulus requirements. The following data is necessary.

(a) The proportion of feed metal available from the feeder which meets
    modulus requirements (C%). Safe values although not necessarily
    the most efficient are:

        33–50% if a Foseco sleeve is being used

        14–16% if it is a live natural feeder (i.e. one through which metal
    has to flow before it reaches the casting cavity in the
    mould)

        10% for other natural feeders.

(b) The shrinkage of the alloy to be cast (S%).
    The shrinkage values for the principal casting alloys are given in
    Table 19.4.

(c) The weight of metal in the feeder under consideration (W_f).
    The total weight of casting (W_c) which can be fed from a feeder of
    weight W_F is

$$W_c = C/100 \times 100/S \times W_F \qquad (11)$$

If the total weight of casting section (W_T) which requires feeding is greater
than W_c then increase the dimensions of the feeder until

Table 19.4 Shrinkage of principal casting alloys

Casting alloy	Shrinkage (%)
Carbon steel	6.0
Alloyed steel	9.0
High alloy steel	10.0
Malleable iron	5.0
Al	8.0
AlCu4Ni2Mg	5.3
AlSi12	3.5
AlSi5Cu2Mg	4.2
AlSi9Mg	3.4
AlSi5Cu1	4.9
AlSi5Cu2	5.2
AlCu4	8.8
AlSi10	5.0
AlSi7NiMg	4.5
AlMg5Si	6.7
AlSi7Cu2Mg	6.5
AlCu5	6.0
AlMg1Si	4.7
AlZn5Mg	4.7
Cu (pure)	4.0
Brass	6.5
Bronze	7.5
Al Bronze	4.0
Sn Bronze	4.5

$$W_T = W_c$$

i.e.

$$W_F = W_T \times 100/C \times S/100 \qquad (12)$$

3. *Calculate the dimensions of the feeder neck*
   (a) Top feeders
       No calculation of feeder neck dimensions are required. If possible, feeder sleeves should be used with breaker cores.
   (b) Side feeders
       The required feeder neck dimensions are obtained from the calculation of the neck modulus (M_N) by applying the ratios:

$$M_c : M_N : M_F = 1.0 : 1.1 : 1.2$$

then use either the endless bar equation (Fig. 19.5) or the diagram in Fig. 19.6.

Grey and ductile irons

1. *Calculate the casting modulus*
Divide the casting into sections and determine the important volume-to-cooling surface area ratios according to Fig. 19.5.

$$\text{Modulus} = \frac{\text{Volume}}{\text{Cooling surface}}$$

$$M_c = V_c/A_c$$

2. *Calculate the feeder modulus*
   (a) The graphite expansion which occurs during solidification of these alloys means that grey/ductile iron castings do not shrink for the full time during which liquid metal is present. The shrinking time (ST) is only a proportion of the total solidification time. This proportion, expressed as a percentage, is determined from the central and left-hand sides of Fig. 19.7, which is used in the following manner:
   (b) Using the known carbon content, move parallel to the carbon line to the appropriate (Si + P) content at point A. Draw a line vertically until it intersects the casting modulus line at point B. Extend a line horizontally to the left until it intersects at point D with the line representing the estimated temperature of the iron in the mould. Read off shrinking time (ST) as a percentage of solidification time.
       Effective feeder modulus is determined by:

$$M_F = M_c \times 1.2\sqrt{ST/100}$$

   (c) Provisional determination of the feeder from feeder sleeve tables published by Foseco companies.

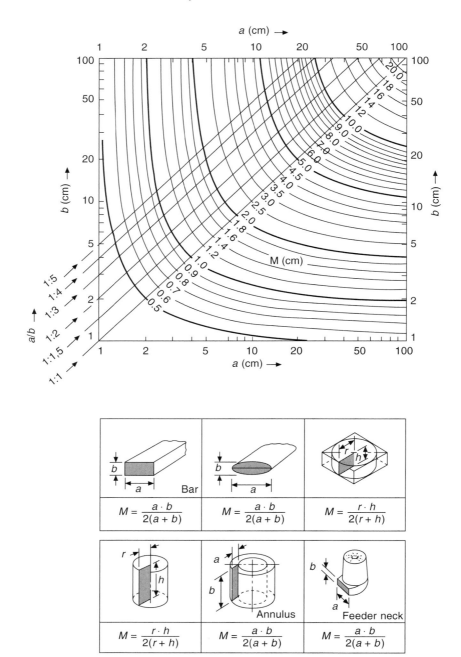

Figure 19.6   *Determination of feeder neck dimensions.*

3. Calculate the feed volume requirement

Feeders, whose dimensions (obtained from the previous pages) satisfy modulus requirements, do not necessarily always satisfy the total feed metal

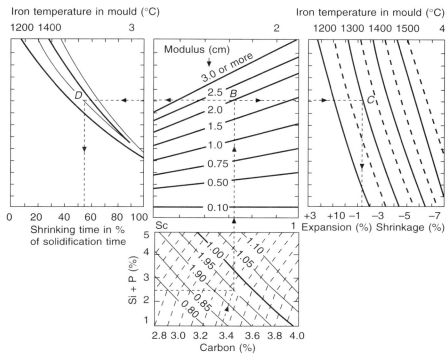

Estimation of shrinkage and shrinking time
from Analysis, casting modulus and metal
temperature in the mould.

Example: 3.35% C, 2.5% Si + P
    Casting modulus: 2.0 cm
    Casting temperature: 1300°C
    Shrinkage: 1.6%
    Shrinking time   55.0%

Figure 19.7 *Determination of shrinkage time and shrinkage for grey and ductile iron castings.*

demand of the casting section. This must always be checked and if the feeder is found to contain insufficient available feed metal the feeder dimensions must be increased. Generally it is preferable to retain the feeder diameter dictated by modulus considerations and increase the feeder height. Sometimes it is more convenient to increase the feeder diameter in place of or as well as height. Never reduce the feeder diameter below that necessary to meet modulus requirements.

The following data is necessary:

(a) The proportion of feed metal available from the feeder which meets modulus requirements (C%). Safe values although not necessarily the most effective are:
33–50% if a Foseco sleeve is being used;

14–16% if it is a live natural feeder (i.e. one through which metal has to flow before it reaches the casting cavity in the mould); 10% for other natural feeders.

(b) The shrinkage of the alloy to be cast. This is determined using the central and right-hand portion of Fig. 19.7. From point B draw a horizontal line until it meets the temperature of the metal in the mould (this must be estimated from the pouring temperature and is usually about 50°C less than the pouring temperature). Read expansion or shrinkage (S) expressed as a percentage on the horizontal axis.

(c) Mould wall movement (*M*%) depends on the hardness of the mould and can vary from zero for hard silicate or resin moulds to 2% for green sand moulds with a mould hardness of 85° (B scale). Add the mould wall movement from the expansion to give a final value of S. If this is positive or zero the casting will not need feeding.

(d) The weight of metal in the feeder under consideration (W_F). The total weight of casting (W_c) which can be fed from a feeder of weight W_F is

$$W_c = C/100 \times 100/S \times W_F \tag{11}$$

If the total weight of casting section (W_t) which requires feeding is greater than W_c then increase the dimensions of the feeder until

$$W_T = W_c$$

i.e.

$$W_F = W_T \times 100/C \times S/100 \tag{12}$$

4. *Calculate the dimensions of the feeder neck*
   (a) Top feeders
       No calculation of feeder neck dimensions are required. If possible feeder sleeves should be used with breaker cores.
   (b) Side feeders
       The required feeder-neck dimensions are obtained from the calculation of the neck modulus (M_N) by applying the ratios:

$$M_c : M_N : M_F = 1.0 : 1.1 \sqrt{ST}/100 : 1.2 \sqrt{ST}/100$$

Then using either the endless bar equation (Fig. 19.5g) or the diagram in Fig. 19.6.

Foseco feeding systems

Introduction

Foseco provides complete feeding systems for foundries, comprising

Sleeves – insulating, exothermic and exothermic-insulating for all metals

Breaker cores – to aid removal of the feeder from the casting
Sleeve/filter units
Application technology – to suit the particular moulds and moulding machine used
Aids to the calculation of feeder requirements.

Range of feeder products

The Foseco range of products comprises:

Product name	Type	Application Alloy	Remarks
KALMIN S	Insulating	Al, iron, steel	extend solidification time by 2.0–2.2 use for hand moulding
KALMINEX 2000	Exothermic-Insulating	iron, steel	extend solidification time by 2.5–2.7
FEEDEX HD	Highly exothermic	iron, steel	small riser-to-casting contact area use where application area is limited or where modulus is the paramount factor
KALMINEX	Insulating and exothermic	iron, steel	for large diameter feeders
KALBORD	Insulating	iron, steel copper-base	in form of jointed mats for large feeders
KALPAD	Insulating	iron, steel copper-base	prefabricated boards or shapes for padding
KAPEX	Insulating	Al, Cu-base iron, steel	prefabricated feeder lids, replace hot-topping compounds
KALPUR	Insulating filter-feeder units for direct pouring	non-ferrous iron, steel	mass production foundries insert sleeves
KALPUR	Insulating and exothermic-insulating filter feeder units for direct pouring	non-ferrous iron, steel	jobbing foundries ram-up sleeves

KALMIN S, KALMINEX 2000, FEEDEX HD, KALMINEX and KALPUR are supplied as pre-fabricated sleeves in a wide range of sizes and shapes, with or without breaker cores.

KALMIN S feeder sleeves

By using a high proportion of light refractory raw materials, a density of 0.45 g/cm^3 is achieved, ensuring highly insulating properties. KALMIN S sleeves are particularly suitable for aluminium, copper-base and iron alloy sand castings since the raw materials used are neutral to both the casting alloys and to the moulding sand. KALMIN S sleeves are supplied as 'parallel conical', 'opposite conical', or cylindrical sleeves for either ram-up application or insertion into cope or drag (Fig. 19.8, Fig. 19.9, Fig. 19.10, Fig. 19.11). They can be applied in the following moulding systems:

Moulding method	KALMIN S feeder sleeve type
Insert sleeves into the turned over cope mould especially in automatic moulding lines	Parallel conical insert sleeves in conjunction with Foseco Sleeve Patterns
Ram-up on the pattern plate in a machine moulding or a hand moulding operation	Parallel conical and/or opposite feeder sleeves, in suitable locating pattern or pin
Insert sleeve in a vertically parted moulding system	Parallel conical and/or opposite conical feeder sleeve in conjunction with appropriate patterns
Insert sleeves into shell moulds	Parallel conical insert and/or opposite conical feeder sleeve
Application of feeder sleeves in a drag mould over a feeder base or bridge core	Parallel conical feeder sleeve as floating sleeve system

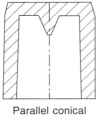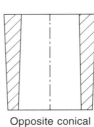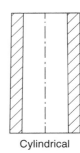

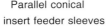

Parallel conical insert feeder sleeves   Opposite conical feeder sleeves   Cylindrical feeder sleeves

Figure 19.8   *KALMIN S feeder sleeve types: Parallel conical insert sleeves; Opposite conical sleeves; Cylindrical sleeves.*

KALMIN S feeder sleeves extend the solidification times by a factor of 2.0–2.2 compared to natural sand feeders of the same size. From these results, modulus extension factors (MEF) of 1.4–1.5 have been calculated. Though KALMIN S feeder sleeves can give more than 33% of their feeder volume to the solidifying casting, it is recommended that a maximum of one-third of the feed metal volume should be fed into the casting so that the residual feeder modulus is adequate in relation to the casting modulus at the end of solidification. For this reason, it is recommended to consider modulus as well as solidification shrinkage in order to determine the correct feeder. Foseco provides tables allowing KALMIN S feeders to be selected with the

Figure 19.9 *Example of sleeve support on a moulding machine pattern plate.*

Figure 19.10 *A section of an insertable KALMINEX sleeve with breaker core attached and a typical sleeve pattern.*

desired modulus, volume (capacity) and dimensions. The KALMIN S feeder sleeve material does not contain any component which may contaminate moulding sand systems and is neutral to the moulding environment.

KALMINEX 2000 feeder sleeves

KALMINEX 2000 is a highly insulating and exothermic feeder sleeve material in the form of prefabricated sleeves for iron and steel casting in the modulus range between 1.0 to 3.2 cm. For light and other non-ferrous alloys the KALMINEX 2000 feeder sleeve material is not recommended. The manufacturing process specifically developed for this unique feeder sleeve material not only ensures a low density of 0.59 g/cm and, therefore, a high grade of insulation but also an additional exothermic reaction peaking at 1600°C. Due to the strength of the KALMINEX 2000 feeder sleeves. they can often be rammed up directly on the pattern plate on many moulding machines.

Figure 19.11   *The KALSERT system in operation. KALMINEX sleeves being inserted into green sand moulds on an automatic moulding line.*

KALMINEX 2000 feeder sleeves can be applied in the following moulding systems:

Application method	KALMINEX 2000 feeder shape
Insert sleeves into the turned over cope mould especially in automatic moulding lines.	Parallel conical insert sleeves in conjunction with FOSECO Sleeve Patterns.
Ram-up on the pattern plate in a machine moulding or a hand moulding system.	Parallel conical insert feeder sleeve, opposite conical feeder sleeve or cylindrical sleeve.
Insert sleeve in a vertically parted moulding system.	Parallel conical insert and/or opposite conical feeder sleeve.
Insert sleeve into shell moulds.	Parallel conical insert and/or opposite conical feeder sleeve.
Application of feeder sleeves in a drag mould over a feeder base or on a bridge core.	Parallel conical feeder sleeve as floating sleeve system.

When determining the solidification times with KALMINEX 2000 feeder sleeves it has been found that they extend the solidification time by a factor of 2.5 to 2.7 compared to a natural sand feeder of the same size. From these results modulus extension factors (MEF) have been calculated between 1.58 to 1.64. Foseco provides tables allowing KALMINEX 2000 feeders to be selected with the desired modulus, volume (capacity) and dimensions.

Under specific practical conditions it has been found that the KALMINEX

2000 feeder sleeves can render 64% of their volume to the solidifying casting. When using feeders with the correct modulus it is necessary to take into account that the modulus of the residual feeder – if more than 33% of the feeder volume is fed into the casting – may not be adequate in relation to the casting modulus towards the end of the solidification. Therefore, it is essential to calculate shrinkage as well as modulus in order to determine the correct feeder sleeve.

FEEDEX HD V-Sleeves

FEEDEX HD V Feeder Sleeves are used for iron and steel casting alloys. FEEDEX HD is a fast-igniting, highly exothermic and pressure resistant feeder sleeve material. The sleeves possess a small feeder volume, a massive wall, but only a small riser-to-casting contact area (Fig. 19.12a). They are, therefore, specially suited for use for 'spot feeding' on casting sections which have a limited feeder sleeve application area. The sleeves are located onto the pattern plate using special locating pins, the majority are supplied with shell-moulded breaker cores. Owing to their small aperture, these breaker cores are not recommended for steel castings, though special larger apertures can be used.

FEEDEX HD V-sleeves are particularly useful for ductile iron castings since with its low volume shrinkage of below 3%, a modulus controlled KALMIN or KALMINEX 2000 feeder will often have more liquid metal than is necessary. The very high modulus and relatively low volume of FEEDEX HDV gives improved yield. In many ductile iron applications, the small breaker core aperture of the feeder means that the feeder is separated from the casting during the shakeout operation and the cleaning cost is reduced.

VS spot feeder sleeves without breaker cores but with a sand wedge between the feeder sleeve and casting are rammed up with the help of a special spring-loaded location pin, Fig. 19.12b. Moulding pressure squeezes the VS sleeve down the pin, compacting the sand. The sleeve incorporates an exothermic locating core to prevent sand entering the feeder cavity during compaction and to heat the feeder neck.

When used in ductile iron applications, it is important to note that the high temperature reached in the highly exothermic feeder can cause residual magnesium in the iron to be oxidised so that there may be a danger of denodularisation on the casting-feeder interface. To avoid this, residual Mg should be greater than 0.045%, inoculation practice should be optimised and thick breaker cores used. Note that when calculating FEEDEX metal volume, only 50% of the capacity should be assumed since part of the metal in the feeder will be lamellar due to oxidation of the Mg in the feeder cavity.

The permeability of FEEDEX sleeves is as high as the moulding sand, so sand system contamination is not a problem.

Figure 19.13a and b shows examples of the use of FEEDEX HD V sleeves on ductile iron castings.

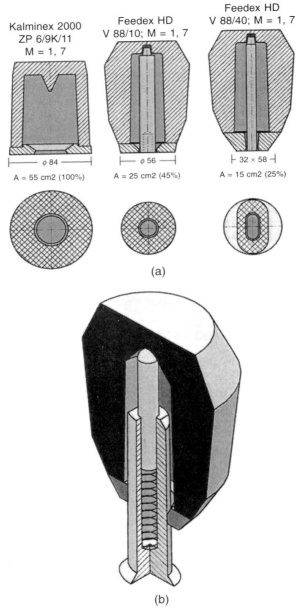

(a)

(b)

Figure 19.12 *(a) The reduction of the feeder contact area which is possible when using the FEEDEX HD V-Sleeve with a suitable breaker core. (b) Sectioned spring loaded steel location pin for ram-up of a V-sleeve without a breaker core.*

KALMINEX Feeder Sleeves

KALMINEX exothermic-insulating feeder sleeves are used for all iron and steel casting alloys. They are supplied with feeder diameters from 80 to

(a)

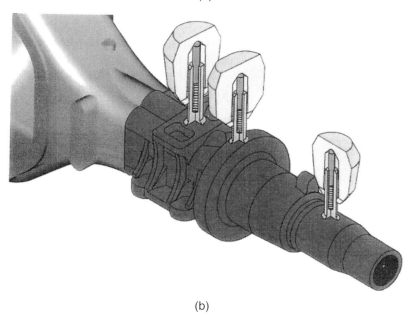

(b)

Figure 19.13 *(a) The application of FEEDEX HD V-Sleeves to a ductile iron turbocharger casting. (b) VS-Spot feeder practice. The point feeding technique makes possible the production of high value, lightweight castings.*

850 mm for the modulus range between 2.4 and 22.0 cm and are suitable for larger sized castings.

The manufacturing process specifically developed for this exothermic-insulating product and the selection of specific raw materials give a total closed pore volume of nearly 50%. The excellent heat insulation resulting from the low density (compared with moulding sand) is enhanced by an exothermic reaction.

When determining the solidification times with KALMINEX Feeder Sleeves it has been found that they extend the solidification time by a factor of 2.0 to 2.4 compared to the natural sand feeders of the same size. From these results modulus extension factors (MEF) of 1.3 to 1.55 have been found.

Under practical conditions it has been found that KALMINEX Feeders when adequately covered with KAPEX lids or a suitable APC (anti-piping compound) may render up to 64% of their contents into the casting. When using feeders with the correct modulus it is necessary to take into account that the modulus of the residual feeder (if more than 33% of the feeder volume is fed into the casting) may not be adequate in relation to the casting modulus towards the end of the solidification. Therefore it is essential to calculate shrinkage as well as modulus when determining the size of the feeder sleeves.

Foseco provides tables allowing KALMINEX feeders to be selected with the desired modulus, volume (capacity) and dimensions. Several different shapes of KALMINEX feeders are available Fig. 19.14a,b,c. Breaker cores are generally made of chromite sand, though they can be produced in silica sand.

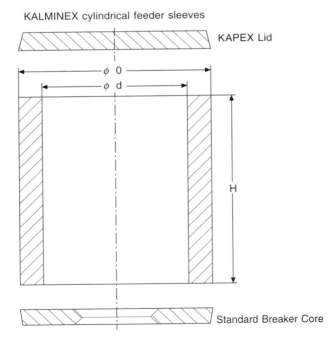

(a)

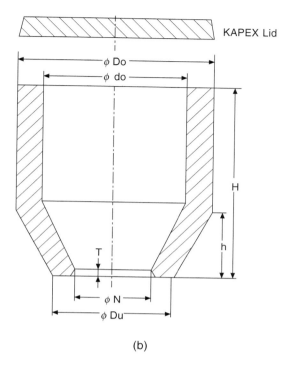

(b)

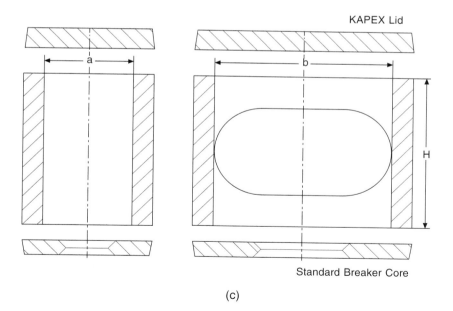

(c)

Figure 19.14 *(a) KALMINEX cylindrical feeder sleeve with breaker core and KAPEX lid. (b) KALMINEX TA sleeve. (c) KALMINEX oval sleeve.*

KALBORD insulating material

Although in theory there is no upper limit of inside diameter for using prefabricated feeder lining shapes for inside diameters above about 500 mm, manufacture, transport and storage become increasingly inconvenient. For this reason Foseco has developed KALBORD flexible insulating material in the form of jointed mats. They can be easily wrapped around a feeder pattern or made up into conventional sleeves as required for the production of insulating feeders for very large steel, iron and copper based alloy castings (Fig. 19.15).

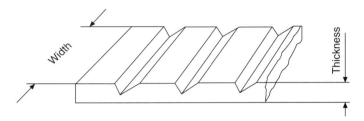

Figure 19.15   *KALBORD jointed mats.*

KALBORD mats are available with 30 mm and 60 mm thickness in widths up to 400 mm and lengths 1020 or 1570 mm. Their excellent flexibility permits the lining of irregular feeder shapes. The mat is most easily separated or shortened with a saw blade.

Produced from high heat insulating materials, 30 mm mats achieve a 1.2 fold and 60 mm mats a 1.3 fold extension of the modulus. It is recommended that KALBORD feeders are covered with FERRUX anti-piping powder.

KALPAD prefabricated boards and shapes

KALPAD has been developed by Foseco to provide a light-weight, highly refractory insulating material to avoid metal padding and to promote directional solidification. If KALPAD insulating shapes are used the desired shape of the casting need not be altered. This increases yield and reduces fettling and machining costs. For this purpose KALPAD is used in copper-base metal and steel foundries, in particular, however, in malleable iron and grey iron mass production.

Owing to a special manufacturing process and the use of alumina mineral fibres KALPAD shapes have a density of 0.45 g/cm^3 with more than 60% of the volume being closed pores which are the reason for the high insulation and refractoriness. During pouring KALPAD produces only negligible fumes and behaves neutrally towards moulding materials and casting metals.

When evaluating solidification times on KALPAD padded casting sections it has been found that they extend the solidification time by a factor of 2.25 to 2.5 compared with conventionally moulded castings. From these results modulus extension factors of 1.5 to 1.58 have been calculated. It is recommended to use a factor of 1.5 if KALPAD shapes of 20 to 25 mm thickness are applied. The dimensions of KALPAD boards and shapes are shown in Fig. 19.16.

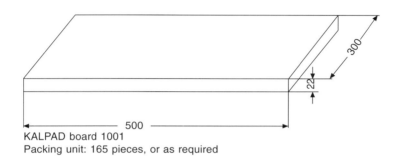

KALPAD board 1001
Packing unit: 165 pieces, or as required

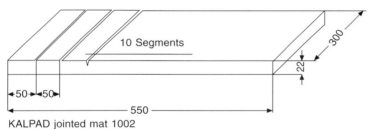

KALPAD jointed mat 1002
Packing unit: 120 pieces, or as required

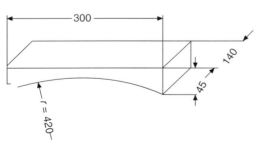

KALPAD pad 1012
Packing unit: 15 pieces

KALPAD 1001 and 1002 are standard types which can be easily sawn to the desired shapes.
KALPAD 1012 is an example of an insulating pad for runner wheels and gear rims.
Shapes made to measure for mass production upon request.

Figure 19.16 *KALPAD prefabricated boards and shapes.*

KAPEX prefabricated feeder lids

KAPEX LD insulating feeder lids (Fig. 19.14a) are an improvement over the hot-topping powders in foundry use, being dust and fume free and giving repeatable feeding results. They can be applied to all feeders either exothermic, insulating or natural. The lids have an insulator density of 0.45 g/cm³ and are purely insulating. Owing to their neutral behaviour towards moulding material and casting metal they are used in light metal and copper-base foundries as well as in high alloy steel foundries. KAPEX LD lids have replaced over 50% of the hot-topping used in Europe for sleeves of diameter between 100 and 350 mm and their use is still growing.

For smaller neckdown feeder sleeves, KALMINEX 2000 exothermic KAPEX lids are also available.

KALPUR filter feeder units

KALPUR filter feeder technology is the shortest way to a perfect casting, see pp. 266, 289. There are two types of KALPUR filter feeder products; one for mass production foundries and the other for jobbing foundries. For mass production, KALPUR insert filter feeder units are inserted and secured into the mould cavity, created by means of suitable KALPUR insert patterns. KALPUR KSET filter feeder units with upper filter location are supplied for horizontally parted automatic moulding machines, Fig. 19.17. For vertically parted moulding lines, KALPUR units with lower filter position are used Fig. 19.18.

For jobbing and simple moulding machine application KALPUR ram-up filter feeders, Fig. 19.19a are used by means of a centering support pattern in combination with a protective bridge pattern Fig. 19.19b.

KALMINEX 2000 ZTAE KALPUR exothermic units and insulating KALMIN STP units are available for ram-up applications.

Use of the KALPUR unit eliminates the conventional gating system and creates the ideal directional solidification condition. The KALPUR filter feeder units must be selected according to their modulus and flow rate with a filter type appropriate to the alloy being cast.

Table 19.5   The KALPUR filter feeder product range

Name	Material	Alloy	Application*	Diameter range (mm)
KALMIN S KSET	Insulating	NF; iron, steel	Horizontal insert	50, 60, 70, 80, 90, 100
KALMIN S KSE	Insulating	NP; iron, steel	Vertical insert	60 and 90
KALMIN 70 STP	Insulating	NF; iron, steel	Jobbing ram-up	50, 70, 90, 125
KALMINEX 2000 ZTAE	Exothermic-insulating	Iron and steel	Jobbing ram-up	100, 120, 150, 180

*NB for aluminium, iron and steel casting KALPUR types are KALPUR AL; KALPUR FE and KALPUR ST respectively in which SIVEX; SEDEX/CELTEX and STELEX filters are integrated.

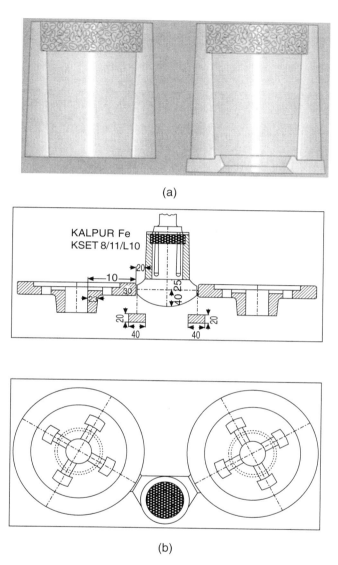

Figure 19.17 *(a) KALPUR insert filter feeder units with upper filter location for horizontally parted automatic moulding lines. Without breaker core for direct pouring through a side feeder. With breaker core for direct pouring through a top feeder. (b) KALPUR Fe Insert side feeder used for feeding two ductile iron case covers.*

Breaker cores

Breaker cores for the reduction of the feeder-to-casting contact area enable feeders to be broken off or knocked off from many types of castings. In the case of very tough casting alloys where it is not possible to simply break off or knock off the feeder, the advantage of using breaker cores lies in the reduction of fettling and grinding costs for the removal of the feeder.

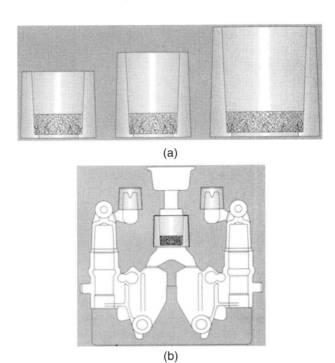

(a)

(b)

Figure 19.18   *(a) KALPUR pouring cup filter-feeder units with lower filter position for vertically parted moulding lines. (b) KALPUR pouring technique used on a DISAMATIC moulding line for casting ductile iron hydraulic arms.*

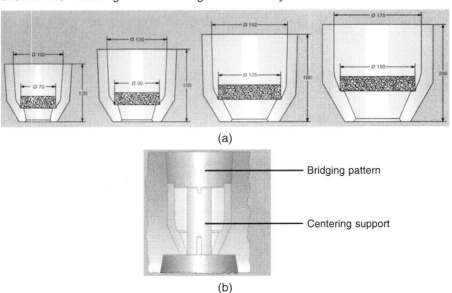

(a)

(b)

Figure 19.19   *(a) KALPUR ram-up filter feeder series for jobbing and simple moulding machine applications. (b) Ram-up principle with fixed centering support and a loose 'bridging' pattern to give the correct height.*

Besides the conventional types of breaker cores based on silica sand (Croning) and chromite sand, special breaker cores with a very small aperture are also in use in repetition iron foundries. These special breaker cores as shown in Table 19.5 are made from highly refractory ceramic.

Table 19.5   Application of breaker cores

Breaker core material	Casting metal	Feeder diameter (mm)
Silica sand	Steel	35–120
Silica sand	Grey, ductile iron, non-ferrous metals	35–300
Ceramic	Grey, ductile iron, non-ferrous metals	40–120
Chromite sand	Steel	80–500
Chromite sand	Grey, ductile iron	200–500

Experience has shown that at least 70% of the breaker core area should be in contact with the casting, in order to level out the temperatures of the metal and the breaker core from the superheat upon or before reaching liquidus.

Some of the standard forms of breaker core available from Foseco are shown in Fig. 19.20. Foseco feeder sleeves can be ordered with or without breaker cores attached.

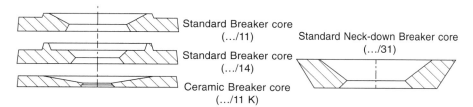

Standard Breaker core
(.../11)

Standard Breaker core
(.../14)

Ceramic Breaker core
(.../11 K)

Standard Neck-down Breaker core
(.../31)

Figure 19.20   *Standard forms of breaker cores.*

The application of feeder sleeves

Ram-up jobbing applications

(1) *Manual moulding*
Sleeves of the correct dimensions are set on the individual pattern in the predetermined location and the mould is rammed around the sleeves. The base of the sleeve should not come into direct contact with the casting but be set on a sand step at least 10 mm thick or the sleeve should be fitted with a breaker core.

(2) On semi-automatic, cold set and slinger moulding lines
If the pattern plate is accessible to the machine operator, the feeder sleeve is located by hand on the pattern plate. To avoid damage during machine moulding, sleeves should be supported by standing them on a pattern dummy or peg at the correct location and having the correct shape and height. Figure 19.9 shows one such arrangement.

Insert sleeves

Often on high pressure, squeeze press or air impact moulding lines, pattern plates are no longer accessible. Foseco recognised these changes in the late 1970s and early 1980s and developed insert sleeve application systems allowing fully automatic machine users to retain all the advantages of employing feeder sleeves without slowing down the moulding cycle.

A prefabricated feeder sleeve with strictly controlled dimensional tolerances is inserted into a cavity formed during the moulding operation by a sleeve pattern of precise dimensions located on the pattern plate (Figs 19.10, 19.11).

The insert sleeve patterns are fixed by screwing them onto the casting pattern and they provide the cavity for the insert sleeve. Owing to the special sealing and wedging system no metal can penetrate behind the inserted sleeves and these cannot fall out from their seat during closing and handling of the mould.

The design of the insert patterns also forms highly insulating air chambers behind the inserted sleeves. This additional insulation increases the moduli of the insert sleeve feeders as follows:

FEEDEX Insert Sleeves	HDP	+ 5 %
KALMINEX 2000 Insert Sleeves	ZP	+ 5 %
KALMIN S Insert Sleeves	KSP	+ 4 %

The insert sleeve patterns have a solid aluminium core with mounting thread and a highly wear resistant resin profile. Insert sleeve patterns are available corresponding to the various types of insert sleeves with and without breaker cores. Both insert sleeves and insert patterns are thus part of an integral system.

Floating feeder sleeves

This is a relatively simple application technique with low feeder sleeve application cost since feeder sleeves are simply placed on the drag parting line. The method is applicable for all moulding machines having a horizontal mould parting line. No problems are encountered regarding strength, spring back etc. of the feeder sleeve. On high pressure moulding lines, more economic and non-polluting insulating KALMIN sleeves can be applied.

A two-part sleeve pattern is used with an integrated feeder base and feeder neck (Fig. 19.21). The drag sleeve pattern is secured on to the drag pattern plate which creates a suitable location and positioning cavity for the corresponding feeder sleeve. The feeder sleeve is simply positioned on this location cavity (Fig. 19.22a). The cavity created by means of the cope sleeve pattern ensures location of the feeder sleeve while closing the mould. After pouring, the feeder sleeve floats along with the liquid metal, secures and seals itself tight into the mould wall cavity created by means of the cope sleeve pattern (Fig. 19.22b).

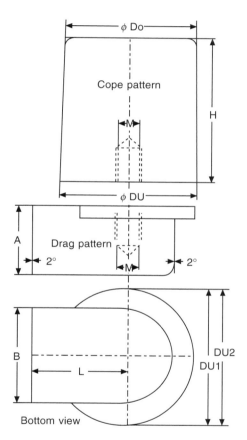

Figure 19.21 *Sleeve pattern for a floating sleeve. The cavity created by means of the cope sleeve pattern ensures location of the feeder sleeve while closing the mould.*

The floating sleeve patterns incorporate maximum feeder neck dimensions applicable to iron castings. For steel, light alloys and non-ferrous alloys, neck modulus can be modified to usual casting modulus equal to neck modulus. For full details, refer to Foseco leaflets.

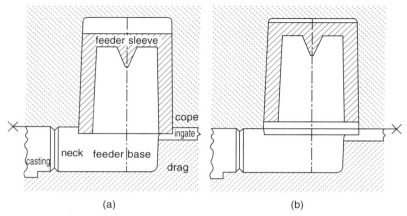

(a)   (b)

Figure 19.22   *Floating sleeve functional principle. After pouring, the feeder sleeve floats along with the liquid metal and seals itself tight into the mould wall cavity.*

Shell mould application

Sleeves may also be inserted into shell moulds. The principle is the same as for green sand moulding, special sleeve patterns are available which form ridges in the sleeve cavity which grip the inserted sleeve but do not damage the mould (Figs 19.23 and 19.24).

Vertical mould or DISA application

Insert sleeves can be applied equally to moulds with a vertical parting, such as those made on the Disamatic moulding machine. The sleeve pattern is divided, but off centre, one part being slightly smaller than the other. The two parts are mounted on opposite sides of the Disamatic pattern plates with the sleeve located in the larger cavity and held in place by the exact vertical fit of the sleeve in the mould. When the mould is closed the second half holds the sleeve fully in position.

Core application

Feeder sleeves may also be inserted into cores. For example, ductile iron hubs are often fed by one or more side feeders located externally to the flange but the most efficient feeding method is by means of a sleeve located in the central core and connected to the casting at the point where the feed metal is really needed. The sleeve is placed into the core print location and fits into the cope mould cavity created by the appropriate sleeve pattern (similar to a floating sleeve pattern); the result is an improvement in yield, cleaning costs and casting soundness (Fig. 19.25).

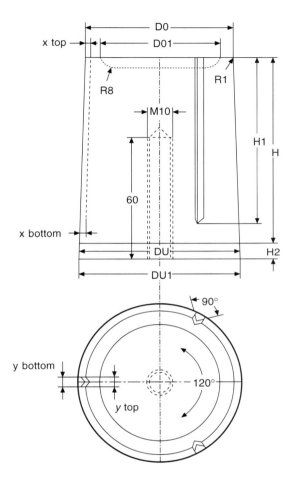

Figure 19.23 *Sleeve pattern for shell mould application.*

Williams cores

The purpose of a Williams core is to prevent the top of the sand feeder solidifying prematurely so that atmospheric puncture can take place on top of the feed metal to promote adequate feeding of the casting. Williams cores are supplied in a range of sizes up to 66 mm diameter D, Fig. 19.26, in FEEDEX exothermic material. KALMIN S and KALMINEX 2000 parallel conical insert sleeves are manufactured with a Williams Wedge incorporated into the design (Fig. 19.8).

Figure 19.24 *A shell mould with feeder sleeves (sectioned for clarity) located in the cavity formed by the shell mould sleeve pattern.*

Figure 19.25 *A ductile iron hub casting using a sleeve located in the central core. The sleeve is placed into the core print location and fits into the cope mould cavity created by the appropriate sleeve pattern.*

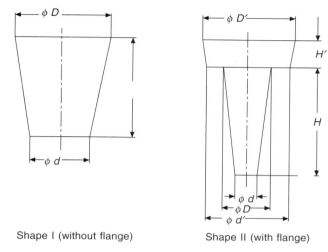

Shape I (without flange)   Shape II (with flange)

Figure 19.26 *Williams cores.*

FERRUX anti-piping compounds for iron and steel castings

The FERRUX range includes anti-piping compounds of all types with reactions in contact with the molten metal which vary from very sensitive, highly exothermic to purely insulating. Described as examples are three grades of FERRUX manufactured in the UK which cover the requirements of the complete range of all ferrous alloys cast in all feeder diameter sizes.

All three grades have an exothermic reaction and one of them, FERRUX 707F, by expanding in use, incorporates the most modern technology. The examples detailed below therefore should only be considered as typical of the types of FERRUX grades and the technology which is available.

The anti-piping compound, pre-weighed and bagged, should be added in the bag to the surface of the metal immediately after pouring has been completed. It is advisable to design the feeder to pour slightly short so that a space can be left between the surface of the metal and the top of the mould. FERRUX will then be contained in this space. The recommended application rate is a layer which has a thickness equivalent to one tenth of the diameter or 25 mm whichever is the greater. If after application the powder is not evenly distributed then the upper surface should be raked flat; normally this will not be found to be necessary.

Note that, for environmental and practical reasons, KAPEX insulating lids are replacing hot-topping compounds in many applications, see p. 322.

FERRUX 16
This is a carbon-free, sensitive, fast reacting exothermic anti-piping compound of high heat output. After the exothermic reaction has ceased, a firm crust

remains on top of the feeder. It is particularly recommended for use on feeders where rapid sculling takes place and where carbon contamination is to be avoided. Feeders where FERRUX 16 is most often employed are in the diameter range 25–200 mm.

FERRUX 101
This is an exothermic anti-piping compound of medium sensitivity. It is ideal for general steel foundry use on feeders of 150 mm diameter and upwards. It may also be used on iron casting feeders where the crust formed after the exothermic reaction has ceased, forms a good insulation against heat losses. The crust can be broken for topping up large castings. The absence of carbonaceous materials in the product ensures that no carbon contamination of the feed metal will occur.

FERRUX 707F
This is a medium sensitivity, exothermic anti-piping compound which expands during its reaction to approximately twice its original volume, to produce a residue of outstanding thermal insulation. In spite of the exothermic reaction, FERRUX 707F is virtually fume free and, in addition, because of the expansion and because of the product's lower density, the original weight of FERRUX 707F which has to be used for effective thermal insulation is usually only about half that of non-expanding grades. The low carbon content of this product will not normally affect metal quality in any significant way. FERRUX 707F is most generally employed for steel and iron feeders of 150 mm diameter and upwards.

Metal-producing top surface covers

THERMEXO is a powdered, exothermic feeding product which reacts on contact with the feeder metal to produce liquid iron at a temperature of about 2000°C. The product is designed for emergencies in case of metal shortage.

Even in the best foundries, occasionally the weight of metal left in the ladle is overestimated and a casting is poured short. The addition of a metal producing compound may save the casting by providing the extra feed metal necessary. In such cases the foundry has nothing to lose by trying a metal producing compound and it is for emergency reasons that every steel foundry should have a stock of THERMEXO.

FEEDOL anti-piping compounds for all non-ferrous alloys

The lower casting temperatures and the differing chemical requirements for non-ferrous alloys necessitates a completely different range of anti-piping compounds than that used on ferrous castings. FEEDOL is the name given to Foseco's range of anti-piping compounds for non-ferrous castings. As an example, two of the principal FEEDOL grades manufactured in the UK are described in detail below.

FEEDOL 1

This is a mildly exothermic mixture suitable for all grades of copper and copper alloys. The formulation does not contain aluminium and there is therefore no risk of contamination where aluminium is an undesirable impurity. After the exothermic reaction has ceased, FEEDOL 1 leaves a powdery residue through which further feeder metal can be poured if necessary. FEEDOL 1 is useful for feeders up to 200 mm diameter. For very large copper based alloy castings such as for example, marine propellers, FERRUX 707F is to be recommended.

FEEDOL 9

This is a very sensitive and strongly exothermic compound recommended for use with aluminium alloys. After the completion of the exothermic reactions the residue forms a rigid insulating crust. FEEDOL 9 is recommended for aluminium alloy feeders of all sizes.

The development of Foseco feeding technology

Fifty years ago, foundries made their own feeder sleeves from sacks of FEEDEX powder supplied by Foseco. In the 1960s Foseco developed a vacuum production technique for the large-scale manufacture of insulating KALMIN and exothermic-insulating KALMINEX sleeves in standardised cylindrical and oval types. At the end of the 1960s the highly exothermic-insulating KALMINEX S sleeve was developed for application on high production moulding machines, placed on the pattern plate and rammed up to form a side feeder. With the increased use in the 1970s of ductile iron for high volume castings and the use of automatic moulding machines which did not allow access to the pattern plate during the moulding cycle, Foseco developed the insert sleeve concept which is still widely used in many high production iron foundries. This concept also made possible the broader application of breaker cores. At the end of the 1990s, more than 60% of all insert sleeves supplied by Foseco have breaker cores.

In the 1990s the V-sleeve concept of a small feeder volume and a small feeder application area allowing 'spot' feeding of castings was developed. By the mid-1990s V-sleeve technology was introduced into most mass production iron foundries. At the end of the 1990s many foundries combine the insert and V-sleeve technologies on fast moulding machines. This allows castings requiring a large number of feeders to be made within the short moulding cycle time available.

The latest VS-Spot feeder is connected to the casting by a very small feeder neck which can be removed with a hammer blow leaving only a small stub to be ground off.

The KALPUR filter feeder unit allows conventional gating systems to be eliminated and creates ideal directional solidification of the casting.

Figure 19.27 illustrates the range of Foseco feeder and filter-feeder products presently available.

KALMINEX cylindrical and oval feedersleeves

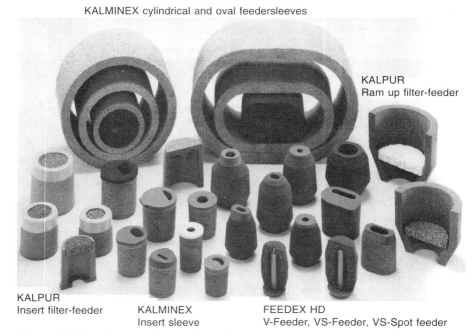

KALPUR
Ram up filter-feeder

KALPUR
Insert filter-feeder   KALMINEX   FEEDEX HD
                              Insert sleeve   V-Feeder, VS-Feeder, VS-Spot feeder

Figure 19.27   *Some of the available Foseco feeding and filter-feeding units. (This figure is reproduced in colour plate section.)*

Aids to the calculation of feeder requirements

Tables

Tables have been drawn up which will convert natural feeders to sleeved feeders for steel castings. No knowledge of methoding is required all that is necessary to know are answers to the following questions.

(a) What are the dimensions of the natural feeder?
(b) What is the weight of the casting section being fed?
(c) What is the alloy composition?
(d) Is the casting with the natural feeder sound?

The tables will do the rest. The conversion, however, is very primitive for if the natural feeder is too large then the sleeved feeder will also be too large and conversely if the natural feeder is too small and causes shrinkage so will the sleeved feeder.

Similarly a simple table has been compiled for ductile iron castings. It is simple to use and requires no expert knowledge of methoding practice. Although in most cases the recommendation if followed will give a suitable feeder sleeve it is not necessarily the optimum size for a given casting section. The table compiled by Foseco in the UK is shown as Table 19.6.

Table 19.6   Feeding guide for ductile iron castings

Weight of casting section (kg)	Sleeve type KSP, ZP, HDP	Sleeve unit No. insert tapered	Feed metal wt. in sleeve (kg)
270.0	16/15	KC3830	19.5
180.0	14/15	KC3826	13.9
130.0	12/15	KC3324	9.8
82.0	10/13	KC3168	5.7
60.0	9/12	KC3596	4.8
37.0	8/11	KC3164	3.0
26.0	7/10	KC3160	2.2
14.0	6/9	KC3156	1.4
8.9	5/8	KC3152	0.92
6.8	4/95	KC3148	0.77
4.5	4/7	KC3144	0.55
1.7	3.5/5	KC3998	0.23

Nomograms

A series of nomograms which relate the casting modulus, which has to be calculated, and the weight of the casting section to a suitable size of feeder sleeve has been developed. Two examples are shown in Fig. 19.28. Such nomograms have distinct disadvantages, they do not take into account many of the variables commonly found in steel foundries, they are however a significant step forward for feeder recommendations can be made without the need to know the original natural feeder dimensions.

FEEDERCALC

FEEDERCALC is a Foseco copyrighted PC computer program which enables the foundry engineer to make rapid, accurate calculations of casting weights, feeder sizes, feeder neck dimensions and feeding distances and to make cost analyses to quickly determine the most cost-effective feeding system for any given casting. Versions of FEEDERCALC have been produced for iron and for steel castings.

The program has been designed by foundrymen for foundrymen and requires only a basic understanding of Windows. Note that the FEEDERCALC program does not replace the experienced methods engineer for it relies on the engineer's skill to interpret drawings, determine feeder locations and accommodate the particular production practice of the engineer's foundry.

The program includes:

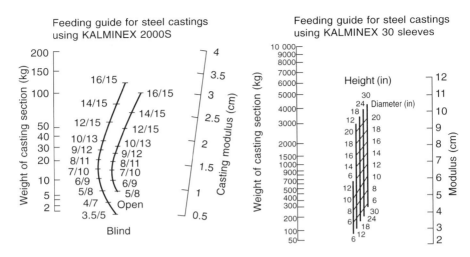

Figure 19.28 Examples of nomograms used to determine suitable feeder sleeve dimensions.

Weight estimation: Combinations of geometric shapes can be easily manipulated to estimate the weight of complex castings or casting sections.

Feeding distance: Feeding distances can be calculated, with and without the use of chills. This allows the rapid determination of the number of feeders required for the casting or casting section.

Feeder size calculation: Accurately calculates the optimum feeder size from a selection based on the complete range of Foseco feeding products. A selection can then be made to meet individual production requirements.

Side-neck calculation: Facilitates the calculation of neck dimensions, weight and fettling areas from the minimum neck modulus supplied in the feeder size calculation.

Cost analysis: Enables the foundry engineer to select not only the optimum but also the most cost effective technique for each casting. Cost data defaults can be used or values specific to the foundry can be entered by the operator.

Casting information: Information relevant to each casting analysed using FEEDERCALC can be saved along with the above data, namely Methods Engineer, Pattern No., Drawing Number etc. Therefore, all records applicable to the casting can be stored and retrieved using FEEDERCALC.

FEEDERCALC for Windows

(Versions 1.0 or later) requires the following computer hardware to run effectively.

100% IBM compatible computer (386 processor minimum; 486 or Pentium recommended)

Minimum 4 MB memory
Hard Disk (minimum 2MB available space)
$3^1/_2$" disk drive
Windows version 3.1 or later
Microsoft Mouse
VGA (or higher) graphics capability
Any printer configured for use with Windows (if hard copy is desired).

The opening Information Screen, (Fig. 19.29) allows the user to save details of a specific casting method and to select the units and material density to be used for the calculation. The default language to be used can also be chosen from English, French, German, Portuguese, Spanish or Italian.

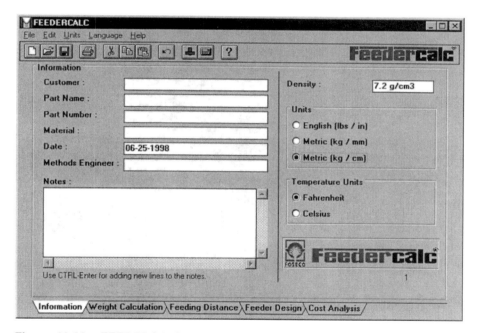

Figure 19.29 *FEEDERCALC Information screen.*

Weight calculation

Single clicking on this tab displays the Weight Calculation screen shown in Fig. 19.30. This screen is used to estimate the weight of a casting or casting section. Weights of complex castings or casting sections are calculated by breaking the casting down into simple shapes, entering dimensions to calculate the weight of each shape, and combining all of the weights to give a total weight for the casting or casting section. (Shape weights may be added to, or subtracted from, the total.)

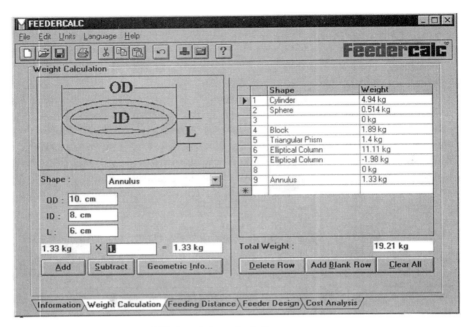

Figure 19.30 *FEEDERCALC weight calculation screen.*

Feeding distance

Once the weight of the casting has been estimated, the Feeding Distance section of the program is used to determine how many feeders are required to feed the casting (see Fig. 19.31).

In the Iron version, the Feeding Distance program is based on a combination of the work of Holzmuller and Wlodawer (whose work relates feeding distance to the amount of graphite present in the alloy), and of the work of R. Heine and of Wallace, Turnball and Merchant relating feeding distance in ductile and grey iron castings to both mould hardness and casting section (modulus). In addition, the calculations in FEEDERCALC have been augmented with empirical casting data based on practical experience. If lower soundness standards are acceptable for the particular casting in question, the user may extend feeding distances shown by the program, based on specific alloy or mould material being used and the user's experience. The program assumes vertical feeding distance to be equal to horizontal feeding distance. Shrinkage or expansion of the alloy under consideration is calculated by inputting total Carbon, Si and P. From this data the program calculates the shrinkage value for the alloy. The mould material pull down list allows the nearest type of mould material to be selected. This is important in ductile irons because mould wall movement must be taken into account.

In the Steel version, the Feeding Distance program is based on horizontal feeding distance path data published by the Steel Founders Society of America and is recommended for low carbon steel sections down to 25 mm thick (to A1 radiographic standard) cast in green sand. If lower soundness standards

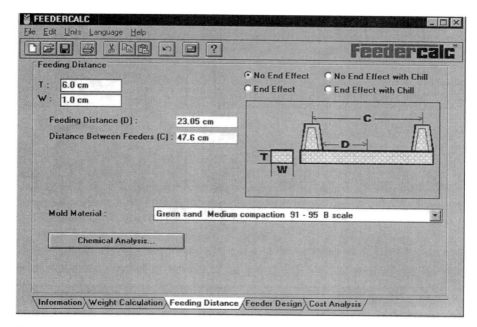

Figure 19.31 *FEEDERCALC feeding distance screen.*

are acceptable for the casting in question, the user may extend feeding distances shown by the program, based on the specific alloy or mould material being used and the user's experience. The program assumes vertical feeding distance to be equal to horizontal feeding distance. The user can now design feeders for the casting by accessing the Feeder Design section of the program. The mould material is selected from a pull down list and the shrinkage of the alloy being used calculated from the type of steel used and the pouring temperature.

Feeder size calculation (Fig. 19.32)

The program determines the optimum size of sleeve required in each Foseco product line which will meet the requirements for the casting section being considered. It also calculates the smallest possible sand feeder which will meet these same conditions, enabling a comparison to be made.
    The calculation is based on:

Feeder position
Casting section weight
The section DIS (Diameter of Inscribed Sphere) in the casting section adjacent to the feeder in question
The number of non-cooling faces
The ingate position and type

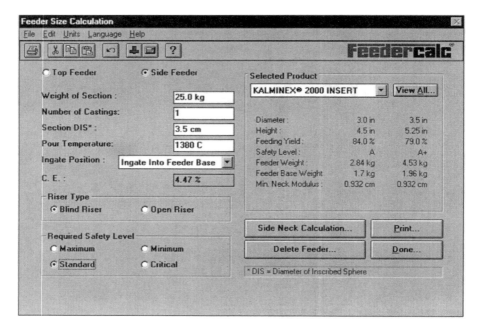

Figure 19.32   *Feeder size calculation.*

The safety margin required
Feed demand from lower section
Superheated core or mould section

Once all the required inputs have been entered into the left hand side of the Feeder Size Calculation screen, the user may select a feeding product from the 'Selected Product' pull down list. Once a product has been selected, the FEEDERCALC program calculates the optimum feeder size for the casting under consideration and this information is displayed beneath the product name. The next size up in the product range is also displayed for comparison purposes.

The Side Neck Calculation button allows the user to calculate the size of a neck used with a side feeder. The neck modulus, provided by the feeder size calculation for the feeder to be used, is automatically transferred to this screen. The program then calculates the range of values which one dimension of a rectangular-section neck would have based on this neck modulus. When the most convenient dimension for the casting being considered is entered, the second dimension is automatically calculated together with the resulting contact area.

Cost analysis

Selecting this option allows the cost of a feeding practice for the casting or castings under consideration to be determined. A comparison of two alternative feeding practices can also be made (for example a practice using sleeves compared against one using sand risers).

By clicking on the 'Cost Analysis' tab, the screen Fig.19.33 appears. Six items of basic foundry cost data are presented. These are average values estimated by Foseco (for the country for which the particular program was designed). The operator may simply accept these (and continue onto the Job Cost button); or the operator may enter actual costs for the foundry and save this data set by selecting the 'New' or 'Save' buttons.

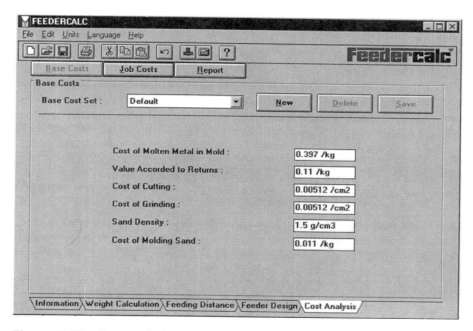

Figure 19.33 *Cost analysis screen.*

The six basic foundry cost elements are:

Cost of molten metal in the mould (currency unit/weight unit)
    This includes costs of raw materials, energy, direct and indirect labour, melting department overheads and any penalty for losses in providing liquid metal to the mould cavity, i.e. melting and pouring.
Value accorded to returns (currency unit/weight unit)
    The value allocated to the metal in feeders, running systems, and scrap castings etc. which are returned for re-melting.
Cost of cutting (feeder removal) (currency unit/area)
    This includes costs of raw materials, energy, direct and indirect labour, cleaning department overheads and penalty for losses in cutting feeders from castings.
Cost of grinding (feeder removal) (currency unit/area)
    This includes costs of raw materials, energy, direct and indirect labour,

cleaning department overheads and penalty for losses for grinding the feeder stub etc. in cleaning castings.

Sand density (weight unit/volume unit)

The density of the moulding medium (sand) when compacted in the mould (typically 1.5 g/cm^2 for silica sand).

Cost of moulding sand (currency unit/weight unit)

This includes the costs of raw materials, energy, direct and indirect labour and overheads associated with preparing the moulding sand (i.e. sand/binder storage, transport and mixing).

Job costs

The Job Costs button accesses the screen of Fig. 19.34 on which details of the casting and feeding are entered allowing two practices to be compared. The final Report screen, Fig.19.35, lists all the costs entered and calculated with the total costs for each practice displayed. Various 'what if?' scenarios may be investigated by changing data on the Base Costs or Job Costs screen.

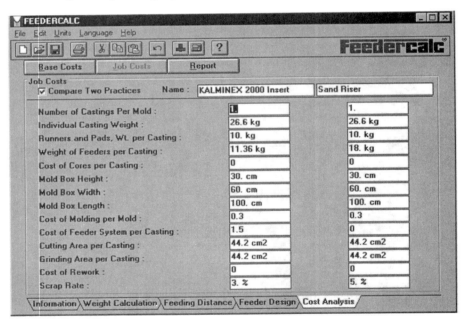

Figure 19.34   *Job costs screen.*

Authorisation

The FEEDERCALC program uses a licensing system to operate the software. If an attempt is made to copy the files, or run the program without proper authorisation, the program will not run and an error message pertaining to

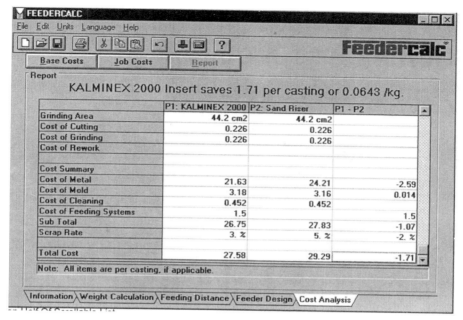

Figure 19.35 *Report screen.*

authorisation will be shown on the screen. Authorisation to run the program is obtained when the Serialization Code for the specific computer on which the program is to be run is notified to Foseco. An Authorisation Code is then provided by Foseco enabling the program to run. The program is set to expire after one year. The user must send Foseco a purchase order to receive the new Authorisation Code.

Chapter 20

Computer simulation of casting processes

Introduction

The purpose of computer modelling of any industrial process is to enable predictions to be made about the effect of adjusting the controls of the process. The casting process is an ideal candidate for modelling. If the process of filling a mould with liquid metal and its subsequent solidification could be accurately and quickly modelled by computer, shrinkage cavities and other potential defects could be predicted. The effect of changing the gating system, the position and size of feeders and even the casting design could be simulated. The casting method could then be optimised before the design and method are finalised, so avoiding expensive and time consuming foundry trials.

Software for the numerical simulation of flow and solidification during casting processes has been available since the mid-1980s. A large number of commercial software packages are now available and they are improving all the time. The modelling of heat flow and solidification of castings is now well advanced. Modelling the filling of castings is more difficult since both turbulent and quiescent flow in complex shaped cavities may be involved. The effects of surface oxide films and bubble entrainment are further complications and it is not easy to check the predictions experimentally in complex moulds.

Solidification modelling

The aim of solidification modelling is to:

Predict the pattern of solidification, indicating where shrinkage cavities and associated defects may arise.
Simulate solidification with the casting in various positions, so that the optimum position may be selected.
Calculate the volumes and weights of all the different materials in the solid model.
Provide a choice of quality levels, allowing for example the highlighting or ignoring of micro-porosity.

Perform over a range of metals, including steel, white iron, grey and ductile iron and non-ferrous metals.

A number of systems are available, they may be divided into two basic types:

Numerical heat flow simulations
Empirical rule based simulations

Numerical based systems, of which the best known is MAGMAsoft, are based on thermophysical data: surface tension, specific heat, viscosity, latent heat, thermal conductivity and heat transfer coefficients of metal, mould and core materials. Mathematical equations are used to calculate heat flow and to predict temperatures within the cooling casting. The complexities of the equations do not allow direct solutions so numerical methods of solving differential equations, finite difference or finite element techniques, must be used. Both require considerable computing power. The time and position where solidification commences is predicted and regions within the casting identified which are likely to become isolated from feed metal. Turbulence during mould filling and convection effects during cooling need to be taken into account.

Empirical rule based systems, such as SOLSTAR, take a model of the casting and its surrounding moulding media divided into small cubic elements. Heat flow and solidification are then modelled by applying iterative rules. In the SOLSTAR method, each element is considered to be at the centre of a Rubik cube of elements with 26 nearest neighbours. The subsequent heat exchange calculations are then carried out in 26 directions taking into account the temperatures and properties of the neighbouring sites. The temperature of a particular site is thus changed step by step resulting in an accurate prediction of thermal history. A liquid site will transform into solid when the site reaches a predetermined value. Metal flow from neighbouring elements is then mathematically simulated to take up the space vacated by shrinkage. When there are no liquid metal elements left to fill the shrinkage cavity a void is created. The system is thus able to predict where shrinkage defects are likely to occur in the casting. By 'calibrating' the rule based system against experimental results, accurate prediction of shrinkage defects is achieved.

While the numerical systems are in principle more precise than rule based systems, at the present time the necessary physical data is not yet available to make numerical systems completely accurate and some 'correction factors' must be introduced.

Rule based systems use standard PC-based computers and are designed to be used by an average foundry engineer. Simulations of freezing of castings are achieved in a fraction of the time needed by numerical models. Numerical systems require more computing power, needing a workstation costing several times more than a PC and requiring a highly trained computer operator.

Both systems are useful to the practical foundryman, not only cutting out trial and error sampling but increasing metal yield, reducing lead times, optimising production methods and improving the accuracy of quotations.

The speed and user friendliness of rule based programs makes them better suited to jobbing foundries where many different castings need to be processed and computation time is an issue. On the other hand, numerical programs demand strong computing power and lengthy processing time, but as better physical data becomes available, they can in principle provide greater accuracy and a closer representation of what actually happens in the mould. They allow temperature profiles after solidification to be modelled so that metallurgical structures can be predicted. They may even take into account the thermal effects that occur during the filling of the mould. Currently such programs are perhaps better suited to research and to study highly engineered, critical components. They are also used for important product development projects, for example automotive castings to be made in very large numbers.

Mould filling simulation

Many of the problems associated with the casting process are related to poor mould filling. Programs have been developed which allow mould filling to be simulated. The aim is to predict how running and gating design affects turbulence in the mould which may trap oxidation products and cause potential defects. MAGMAsoft (and others) have developed such programs which allow the visualisation and animation of the movement of the melt surface during filling. Current limitations are system time, it can take several days to simulate the flow in a mould, and the lack of good data on surface tension, viscosity etc. Several laboratories around the world are generating the thermo-physical data needed to improve the simulations.

The SOLSTAR solidification program

SOLSTAR is used by a large number of foundries, service bureaux and educational establishments. The breakdown of usage (in 1994) is:

Steel foundries	36%
Iron foundries	26%
Non-ferrous foundries	10%
Education & Service	27%

The procedures for carrying out a SOLSTAR analysis are:

(1) Using the casting drawing, determine model scale and element size.
(2) Make the solid model of the casting.
(3) Make the solid model of the proposed production method (feeders,

chills, insulators etc.). Use the program's own feeder-size calculator if required.

(4) Carry out thermal analysis to establish the order of solidification.
(5) Carry out solidification simulation to a set quality standard, for the selected alloy incorporating shrinkage percentage, ingate effects etc. This results in the model being changed to the predicted final shape (internal and external) of the casting showing size, shape and location of shrinkage cavities in casting and feeders.
(6) Examine the predicted shrinkage (the equivalent of non-destructive testing) by viewing and plotting of 3D 'X-rays' and sections of the model in 2D slices or 3D sections and relating predicted defects to solidification contours and required quality standards.
(7) If the predicted defects do not meet the required quality standard, develop an improved production method and repeat the procedures.

These trial-and-error sampling procedures can be carried out very rapidly, allowing the operator to indulge in any number of 'what-if' experiments.

Solid modelling

The first stage in any solidification simulation is to create a three-dimensional model of the component with its associated method. This will often take the greatest proportion of time, as much as 70%. The SOLSTAR program has its own solid modeller/mesh generator capable of modelling the most complex casting shapes. Depending on the computer hardware specification, models can contain up to 256 million elements but most models use between 2 and 64 million elements. Figure 20.1 shows a solid model of a 350 kg steel valve casting containing 40 million elements produced in less than 3 hours.

It is possible to transfer 3D models from any other CAD system using Stereolithography STL files created by them. These models can then be manipulated within the solidification software so that the method can be added.

Thermal analysis

The thermal analysis calculates the simulated heat flow between the elements of the solid model which gives a 'thermal picture' of the conditions prevailing at a specific point in time. SOLSTAR's thermal analysis simulates 'heat flow' in 26 directions, with each cuboid element of metal, mould, chill etc. being the equivalent of the centre block of a 'Rubik' cube (27 cubes).

SOLSTAR uses the thermal analysis to store details of the solidification order of each element of the casting and feeding system. Figure 20.2 shows a 'thermal' illustration of a section through the steel valve casting (in black and white on p. 349 and reproduced in colour in plate section). This is produced in colours showing 'solidification contours' of the metal from the

Figure 20.1 *Solid model of a valve casting utilising 40 million elements. (This figure is reproduced in colour in plate section.)*

Figure 20.2 *Solidification contours for the lower section of the valve casting. (This figure is reproduced in colour in plate section.)*

end of pouring to the end of solidification. The thermal analysis for the steel valve involves approximately 200 billion 'heat exchanges' between adjacent elements of the model and was calculated in 22 minutes using a 266 MHz computer.

Solidification simulation

After the thermal analysis is completed, each metal element of the model is allocated an order of solidification. SOLSTAR then carries out a solidification simulation of the metal elements, by solidifying them in the predetermined order. During this simulation several things are happening:

(1) The solidified elements are assumed to have increased in density, accompanied by a loss of volume.
(2) This loss (liquid shrinkage) is calculated by multiplying the number of solidifying elements by the input shrinkage factor for the alloy.
(3) The software calculates (according to the alloy and the required quality standard) whether this shrinkage will manifest itself in the form of a cavity and, if so, how big the cavity will be.
(4) The resultant cavity is placed in the remaining liquid of the section of which it is a part. Where it resides in this remaining liquid depends on the type of alloy.

(5) The program continually checks the linkages between the remaining liquid metal.
(6) Metal elements are continually 'flowing' through the liquid paths to replace volume loss during solidification, so accurate tracing of these paths is critical to the program.

At the end of the solidification simulation the model represents the casting (and feeders) at the 'shake-out' stage of production. Figure 20.3 is an 'X-ray' plot of the final model showing all the shrinkage cavities predicted to be outside of the requested quality standard. A Class 2 solidification simulation for this valve took 26 minutes using a 266MHz computer.

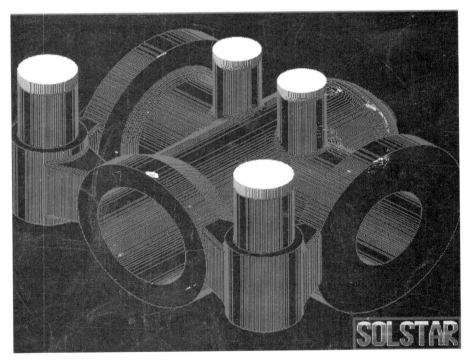

Figure 20.3 *An 'X-ray' plot showing predicted shrinkage cavities. (This figure is reproduced in colour in plate section.)*

During the solidification simulation, the effect of varying

moulding position,
ingate position,
mould materials, chills, insulating and exothermic materials,

can be modelled, allowing the optimum method of making the casting to be predicted.

Feeder size and weight calculations

The program can calculate the volume, weight and surface area of each material in any selected part of the model, so that casting yield is readily obtained. The program can also calculate feeder sizes for steel and ductile iron, giving a selection of feeder options (height/diameter ratios, sand or insulating/exothermic materials). This data allows more accurate estimation of production costs to be made.

Cost benefits of solidification simulation

Figure 20.4 shows a simplified trialling process for making a cast component, the cycle may need repeating several times before product of acceptable quality is made. Solidification simulation software such as SOLSTAR can be used to electronically sample the method and cut down on the number of actual trials made (Table 20.1).

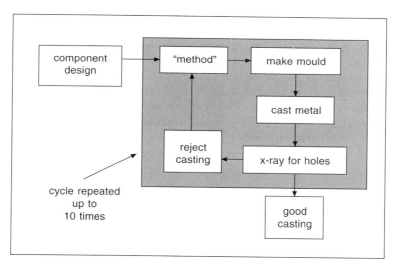

Figure 20.4 *The trialling process for making a cast component of acceptable quality.*

Table 20.1 Pre- and post SOLSTAR methoding accuracy in a steel foundry

Trial methods	Before SOLSTAR	Using SOLSTAR
Right first time	50%	99%
Two attempts	85%	100%
Three attempts	98%	

(From M. Jolly, *Foundryman* Jan. 1994 p. 11)

Each trial carries a cost in materials and manpower but, more important still is the effect on lead time for producing a component. For new jobs in a foundry, where solidification simulation is practised as standard procedure, components can be modelled and methoded at a rate of 10 each week per method engineer, and most of these methods will be right first time (Table 20.2).

Table 20.2   Probability of right first time reported by SOLSTAR users

Steel	99%
Iron	90–99%
Non-ferrous (short freezing range alloys)	99%
Non-ferrous (long freezing range alloys)	85–95%

Conclusions

While this chapter has concentrated on the benefits of the SOLSTAR program to foundries, there are many other rule based programs which can give similar good results. It is important to remember that programs such as these, as well as the more comprehensive numerical packages, must be used by skilled foundrymen who are able to interpret the results to achieve practical solutions in the special circumstances of their own foundries. It must also be remembered that, however good the simulation is, no foundry has complete control over its manufacturing processes so some variation in the end result is inevitable and safety factors must be built in to the design of the casting being made.

Index

R   }
E